Evolve

Sharon Daniels

Embracing the Future of Shared Intelligence

Evolve

Leveraging Artificial Intelligence and Natural Language

Technologies to Enable Superhuman Capabilities

Forbes | Books

Published by Forbes Books, Charleston, South Carolina.
An imprint of Advantage Media Group.

Forbes Books is a registered trademark, and the Forbes Books colophon is a trademark of Forbes Media, LLC.

Printed in the United States of America.

10 9 8 7 6 5 4 3 2 1

ISBN: 978-1-95086-346-4 (Hardcover)
ISBN: 978-1-95588-481-5 (eBook)

Library of Congress Control Number: 2025914881

Cover and Layout design by Matthew Morse.

This custom publication is intended to provide accurate information and the opinions of the author in regard to the subject matter covered. It is sold with the understanding that the publisher, Forbes Books, is not engaged in rendering legal, financial, or professional services of any kind. If legal advice or other expert assistance is required, the reader is advised to seek the services of a competent professional.

Since 1917, Forbes has remained steadfast in its mission to serve as the defining voice of entrepreneurial capitalism. Forbes Books, launched in 2016 through a partnership with Advantage Media, furthers that aim by helping business and thought leaders bring their stories, passion, and knowledge to the forefront in custom books. Opinions expressed by Forbes Books authors are their own. To be considered for publication, please visit **books.Forbes.com**.

09-14-2025 5:28

To my family and partners who have inspired me and supported me to think big. I am who I am today because of your influence, and our commitment to each other, and to innovation. We have a truly special collective passion for the art of technology, and endless enthusiasm for ways in which we can impact a purpose-driven world. I am most grateful for the incredible people around me who seem to toggle both sides of their brain on demand. To the global team and shareholders at Arria, it is an honor to work alongside such intelligence and creativity. Being part of this team allows me to learn, grow, share, and evolve in ways I never imagined.

Acknowledgments

To all the global innovators who have influenced the dawn of a new era called generative AI.

Natural language is uniquely human. Every organization in the world, no matter the size, location, or industry, depends on language to communicate, comprehend, and provide direction. Automating this uniquely human ability offers unprecedented value to organizations. Giving the power of language to the machine will change the way humans interact with data.

The development and application of generative AI technology require collaboration across multiple disciplines, including computer science, psychology, ethics, linguistics, and more. Interdisciplinary collaboration impels innovation while helping to ensure that generative AI is developed and used in a responsible and impactful way.

Language models are fundamental to bridging technology with humanity to reshape and enhance our businesses, our lives, our world. The best is yet to come.

Contents

PART I

EVOLVING THE LANGUAGE

PART II

EVOLVING THE MYTHS

PART III

EVOLVING THE TECHNOLOGY

A language is not just words. It's a culture, a tradition, a unification of a community, a whole history that creates what a community is. It's all embodied in a language.

—NOAM CHOMSKY

Foreword

BY ADAM BRYANT

In mid-April of 2020, roughly two months into the pandemic, my wife and I were holed up in our New York City apartment, and, like everyone else, we were deeply worried about COVID-19, particularly in those early weeks when the science was less clear on how the virus was transmitted. With sirens blaring constantly outside and many people fleeing the city, we went into lockdown, buying masks and wiping down our groceries and struggling to keep an even keel. All the uncertainty—about the virus itself and how long the pandemic would last—felt overwhelming at times.

Thankfully, work provided some distraction. I'm the managing director of the ExCo Group, an executive mentoring and leadership development firm, and one of my roles at the firm is to use the skills I developed over thirty years as a journalist to do interviews with CEOs and other senior executives about their key leadership lessons and insights. Over the last dozen years, I've interviewed more than seven hundred leaders, starting with the Corner Office series I created in *The New York Times* in 2009, and now on LinkedIn, where my various series have more than 175,000 subscribers.

While my focus had always been around timeless insights on leadership, I pivoted quickly when the pandemic started, and I began interviewing leaders about the new playbook for leadership they were writing in real time during the pandemic to help them navigate this historic crisis.

Luckily for me, one of them was Sharon Daniels, the CEO of Arria. I first met Sharon in 2017, when I interviewed her for my Corner Office series in *The New York Times*. It was a powerful conversation, filled with memorable stories and insights. I have thought often about this particular passage from that first interview, as it's a crucial reminder about self-awareness:

> It's important to take the time to have some self-awareness about what has influenced you, and what you stand for, and being conscious of what's actually going on. One of my life principles is inspect your motivation about why you're doing something. You want to do a self-check that you're on the right path and that your actions aren't driven by ego. That way, you can calibrate your actions and how you communicate around the right purpose. Otherwise, it's easy to get caught up in the moment.

For our second interview, shortly after the pandemic started, Sharon and I connected over Zoom, and she shared an insight that led to a wholesale change in my approach for navigating the pandemic. Here's the passage from our conversation that was burned into my brain the moment she said it:

> This crisis is like a reset if you take the opportunity to look at things that way—for the work we do and in our personal lives... It's an important process that we are able to go through as human beings. There's disruption, and it causes motion of some sort. But then we figure out how to level out and find that comfort zone again, which is one of the beautiful things about the human spirit. But without disruption, there is no motion. So people need to focus on the idea that this disruption is causing motion, and I want to make sure it's bringing us forward.

In that moment, Sharon had clarified the challenge for me: How do I take advantage of all the disruption? I got up off the mat, and with my head now in the right place, I decided to make the most of the coming months, even with all the uncertainty ahead. Since all business travel had stopped, I was now more in control of my schedule. A more predictable routine helps me when I'm writing, and I used the time to finish my most recent book, *The CEO Test*. I started running most days and dropped the dozen pounds that had crept up on me when I was traveling more for business.

It has been a reminder of one of the many paradoxes of life. While we often crave a degree of sameness and certainty (and yes, comfort), the stress and uncertainty of disruption can also lead to important moments of growth and stronger muscles for those times that require resilience.

And that is a central theme that Sharon brings to life in her provocative and enlightening book, fittingly called *Evolve: Leveraging Artificial Intelligence and Natural Language Technologies to Enable Superhuman Capabilities*. As you will discover in the pages ahead, she provides a sweeping history of how humans learned to communicate, along with a penetrating look into how artificial intelligence will build and transform the science of natural language generation, which is Arria's specialty.

At its essence, it is the science of turning raw data into an easy-to-understand narrative. And Sharon has applied that art form of simplifying complexity to this rich topic, illuminating an essential field of study that all human beings share: communications.

—ADAM BRYANT
Managing Director, ExCo Group

An Unwavering Believer

In 2002, I was on a business trip in Zurich when I learned of a cooperative community in the remote hills of Switzerland where citizens had allegedly built a free-energy device. Not knowing much more about it—but always eager to witness innovation at work—my business partner and I traveled two hours to the outskirts of Linden in the Swiss Alps. It was a clear, sunny day, and when we arrived, we were greeted by a man named Francis who explained the tenants of their 140-member community: self-sufficiency and surrender. It was easy to see from the windmills that dotted the landscape and the carefully planted rows of crops that the members had worked hard to ensure their sustainability. In addition, each member had forgone their worldly possessions and now lived humbly off the land and their own resources. When we asked about the free energy they claimed to harness, Francis shared that the mystery of its operation remains a secret.

In truth, many scientists say this type of free-energy machine is impossible, as it violates some laws of thermodynamics—and yet, I once witnessed something similar myself! A dear friend and inventor produced a similar prototype machine that successfully generated power from the spinning of his carefully crafted wheel of sculpted metal and detailed mechanisms. As I spun the wheel in a dark room, enough power was pulled from thin air and transferred into my hand, and I was able to turn on the light bulb, simply by touching it. It seemed miraculous. Whether the technology behind it was sound or not, I was able to power the bulb with my touch.

I thought of this miraculous moment as Francis walked us around his community. Whether or not their discovery was sound, I admired the community's philosophy as Francis explained it: "Our energy discovery originated from our own spiritual quest to discover what enables humans to always get back to center. We can be so disrupted—whether by loss, financial hardship, illness—but we can return to our emotional center. Why is that?" I told him his words reminded me of shaking a snow globe. At first, the globe is filled with chaotic particles, but with time the particles settle, and the globe clears. "That's it!" he declared. "It was on our journey to solve a spiritual quandary that we discovered a universal answer: Without disruption, there is no motion. Understanding the first law of motion on a human level helped us envision it on a technological level. If we could keep the disruption going, we could also keep the energy flowing."

It's the law of perpetual motion that allowed this community to embrace literal and figurative disruptions—or "unbalanced external forces" as the law states—rather than fear them. The scientific reality is that, whether we are comfortable with change or not, it is the natural order of the universe. We can spend our human energies trying to make things last beyond this natural order, but it is futile.

I was struck by this community's willingness to embrace disruptions. Francis and the other community members were able to innovate because they embraced a shared vision that movement starts with disruption, whether it be in the universe or in humans' own lives. As I have witnessed over my years in the technology industry, so many great ideas don't make it to the world because they need a group of pioneers who can hold the vision, believe in it, and see it through. As cultural anthropologist Margaret Mead, who was best known for her study of native people of the Pacific, once said, "Never doubt that a small group of thoughtful, committed citizens can change the world. Indeed, it is the only thing that ever has." Not everybody can hold the vision. It needs a group of unwavering believers.

DIGGING DEEPER INTO PERPETUAL MOTION

Sir Isaac Newton penned a famous law in the late 1600s: "Every body remains in a state of constant velocity unless acted upon by an external unbalanced force." Think of being on a playground swing: Once you start swinging and pumping your legs, you can, in theory, swing forever. Once your legs become too tired to continue pumping, however, you remove the energy, and the swing slows. Like the swing, perpetual motion requires an initial force and then a sustaining force to keep it going.

Perhaps one of the best examples of a perpetual motion is the moon. It has been traveling around Earth for a very long time—ever since Earth crashed into a Mars-sized object—and even though its motion seems unchanged, advanced instruments show that it gets farther from Earth by approximately two centimeters each year. Why? Because even in space there are unbalanced external forces at work.

Do scientists think a perpetual motion machine is possible? No, though they suspect some machines can approximate the process. According to Dan Frey, an associate professor of mechanical engineering and engineering systems, "The laws of physics indicate that perpetual motion would occur if there were no external unbalanced forces, but there are. Only by engineering a solution by which an object in motion can consume some store of energy or gather energy from an external source can we approximate perpetual motion."[1]

Though I don't pretend to understand exactly what Francis and the others discovered in those beautiful hills of Switzerland, our conversation changed my perspective and has informed my personal philosophy ever since: Whether we realize it at the time, and whether we like it or not, without disruption there is no motion. When we remain comfortable, we also remain static and inert. I am now an unwavering believer that when we are disrupted—whether internally or externally and whether on a personal level or on a larger, cultural level—we have the chance to grow. In the same way, if our technology never changes, the human race misses an opportunity to evolve.

We have all been disrupted—some are minor disruptions like a flat tire or misplaced keys, and others are more profound, like the loss of a job or the death of a loved one. In 2020, however, all of humanity was collectively disrupted. Almost overnight, news of a global pandemic spread, and we went from deciding where to eat and what to do for

---

1   Jason M. Rubin, "Ask an Engineer," October 4, 2011, https://engineering.mit.edu/engage/ask-an-engineer/is-it-possible-to-construct-a-perpetual-motion-machine/.

the weekend to deciding if it was safe to return to work and send our children to school.

As with most disruptions, we may not see the potential for growth right away. With the pandemic, we were more concerned with our families' health and wellness. We were concerned about our jobs and livelihoods. We didn't need to be reminded of our opportunities for evolution as much as we needed information we could trust and understand. Instead, we were reliant on scientists, charts, and seas of spreadsheets. There was data available, but we didn't know what it meant. We knew that those statistics and graphs meant things that were dear to us and held the answers we were desperately seeking, but how could we use them to benefit our families? None of us could become infectious disease experts overnight (though that didn't stop many of us from trying). How could a disruption of the entire human race evolve us as a species? And how could we hold the vision and be unwavering believers that even this global disruption held the potential for miraculous transformations?

The sheer flood of data and information we were seeing daily about the pandemic was impossible to process. The challenge wasn't to ensure that data scientists were informed; it was to bring collective understanding to every person on the planet. People wanted to understand what was happening, particularly in the areas where they lived, but if that information was too confusing and complicated, people would remain confused, scared, and inert, wondering what to do, how to help, or how to keep their families safe. Rather than evolving, we would stay static, too paralyzed to push ourselves, and thus miss an opportunity to renovate humanity.

Suddenly the stakes of our daily decisions grew, as did the consequences of choosing poorly. Researchers studying human decision-making have found that many of our poor decisions are linked to "the presence of too much information."[2] We all experienced this firsthand. The same researchers claim that one successful strategy for decision-making is to "replace intuition with formal analytic processes—taking into account data on all known variables," which has been shown to significantly improve decisions.[3] There was just one problem: This ideal form of decision-making wasn't feasible because it was humanly impossible to analyze all the pandemic data, especially considering that studies suggest humans can only hold between one and four items of information in our minds at a time.[4] What we quickly learned was that the value of information was not in having access to it; it was in understanding it. We needed what we all need when we are making decisions: facts with context. And we needed it communicated to us in a way we could understand.

The best way to comprehend something is to have it relayed in our own language, which is the most sophisticated medium for information transfer among humans. As the CEO of Arria, which specializes in a form of artificial intelligence (AI) known as natural language technology (NLT), I understood that our technology could empower users by transforming raw data—incomprehensible to nonexperts—into an easy-to-understand narrative that would offer the peace people sought. To this end, Arria partnered its NLT with Microsoft Power BI Visual Analytics technology to convert John Hopkins's data into

---

2   Katherine L. Milkman, Dolly Chugh, and Max H. Bazerman, "How Can Decision Making Be Improved?," University of Pennsylvania Scholarly Commons, July 2009, https://repository.upenn.edu/cgi/viewcontent.cgi?article=1306&context=oid_papers.

3   Milkman et al., "How Can Decision Making Be Improved?"

4   "Understanding the Mechanisms of Language Comprehension," *Research Features*, November 22, 2016, https://researchfeatures.com/2016/11/22/understanding-mechanisms-language-comprehension/.

language—called the COVID-19 Live Dashboard, a web-based dashboard that gave people up-to-date information about the virus.

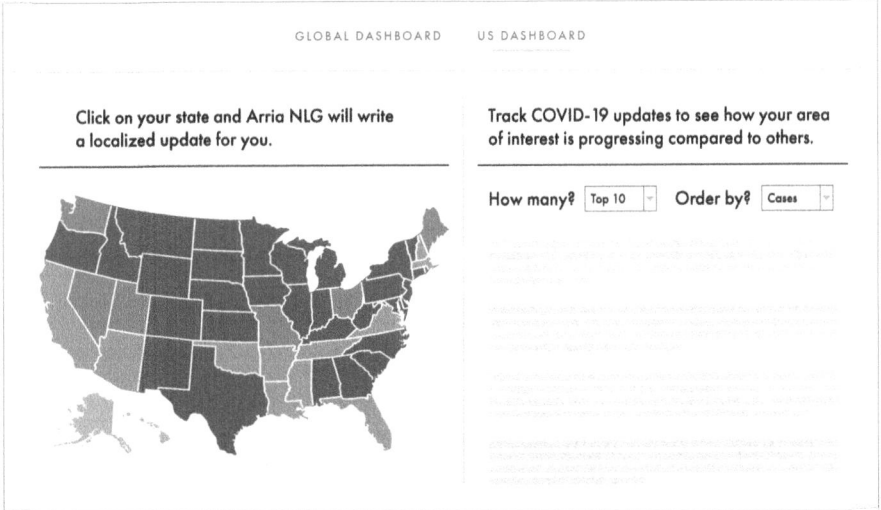

GLOBAL DASHBOARD   US DASHBOARD

Click on your state and Arria NLG will write a localized update for you.

Track COVID-19 updates to see how your area of interest is progressing compared to others.

How many? Top 10   Order by? Cases

Figure 0.1. The COVID-19 live dashboard turned Johns Hopkins's data into narratives that were easy to read and understand. This became a daily go-to, keeping people informed, helping to educate the masses, and closing the knowledge gap. Arria removed the need for data analytics and democratized data understanding in a time of need.

The scientific foundation for Arria's natural language technology is based on more than thirty years of research and development by our founding scientists at the University of Aberdeen, Scotland. The university's global leadership in the field of computational linguistics stems directly from their dedication to the evolution of language for over 520 years. Despite its roots in the past, natural language generation is an innovative AI technology that translates data into linguistic data stories. An important outcome of creating linguistic data stories is that they can be directly interpreted. No more staying up all night to decide whether or not a trip to the grocery store was worth the risk. No more waiting for an expert to tell us whether or not schools were safe for our children because linguistic data stories democratized the data riches to every one of us. We all could have access to the data that we needed, right when we needed it. Simply put, natural language software transforms data into language, but, on a larger scale, adding a narrative to complex data aids understanding, demystifies the unknown, and democratizes data so that we can make real-time decisions for the benefit of ourselves, our families, and our communities.

The law of perpetual motion maintains that *without disruption, there is no motion*. How does a pandemic aid our evolution? How was its disruption benefitting humanity? It certainly didn't feel like an evolutionary leap at the time, but as I thought back to watching my hand power a light bulb, I was reminded that the universe's natural order is change. If we fought external forces in our own lives—and even in our sciences and innovations—we were risking inertia. Whether it was embraced by humans or not, whether we wanted it or not, we were on the cusp of evolutionary change. And it seemed that Arria's AI technology was positioned to lead the way.

Though it might sound counterintuitive, the reality is that AI technology can aid human evolution. As we experienced during the global pandemic, it has the power to make us a smarter, more informed world where expertise is demystified and democratized. With it, we are like the Swiss community—more self-sufficient by surrendering the need to be experts. Like those community members, I stand as an unwavering believer that disruptions lead to advanced technologies that can propel human evolution. Without technology, we remain static, overwhelmed, and uninformed; with it, we restore equality consciousness and are freed to do more, be more. The real technological breakthrough of the last decade is the understanding that humanity is empowered by technology, and technology is empowered by humans.

At Arria, we capture the symbiotic relationship between humans (Hu) and artificial intelligence. Hu + AI = superhuman.

In the following pages, we will discover the amazing evolution of language and technology, not only to better understand its importance in our lives but to invite others to hold the vision for a future where technology is not feared but rather embraced as a tool for superhuman growth. In part I, we will look at the history of language, then debunk some of the most common myths and misconceptions about AI in part II, then move into part III with a look at both the miraculous work being done at Arria right now and the even more miraculous work that lies ahead. As we will learn, natural language generative AI isn't just some new kind of tech. It's a technology with an origin story from more than half a millennia ago! And it is literally transforming, and potentially saving, the world one unwavering believer at a time.

PART I:

Evolving the Language

CHAPTER 1

In the Beginning

When I entered the underground vault beneath the University of Aberdeen in Scotland, the first thing I noticed was the smell—a pungent mix of old leather and something unnamable, ancient. The university's collection of language artifacts was encased in glass and lined the space in chronological order starting from the third century BC. As I walked the room, I traveled through time, charting how language had evolved alongside humans. The earliest works, which predated language, were pictures etched carefully onto clay, papyrus, and wood. I imagined the ancient humans who took such care with the intricacies of their work. I imagined them penning parchment by firelight, binding them in leather they had cured themselves, folding the pages carefully so that their words and thoughts, their knowledge—their very existence—could be retained and shared, boundless and eternal.

Humans' greatest power is our ability to communicate. For most of the last forty thousand years, humans have been sharing our stories about ourselves, our world, and our evolving understanding of its force on us.

As one of the most famous anthropologists of her time, Margaret Mead conducted extensive field research of seven primal cultures between the 1920s and 1960s and remarked that every culture she looked at, despite their many differences, all shared a "cosmic sense"—a need to know how their individual story related to the cosmos as a whole. She discovered they achieved this understanding through their tribal stories, which they then passed on to younger members.

For renowned mythologist Joseph Campbell, who explored the origins of myth by studying the folktales of primitive tribes, he recognized the folktale was "the primer of the picture language of the soul."[5] These early stories from tribes around the world were not just series of sounds. They were ways for outsiders, like Mead and Campbell, to access the cultures' values and norms and, at the same time, the psyches of the people telling them.

There are almost seven thousand languages spoken around the world, and no matter who we are or where we are, language is one of the most universally important aspects of our lives. Language is what builds and cultivates our relationships, allows us to collaborate on tasks and collect the information required for innovations and advancements. Research has shown that humans speak, on average, sixteen thousand words a day.[6] Though we often take these small symbols for granted when we scrawl a quick grocery list, text a spouse about dinner plans, file a client folder to access later, or sign with a deaf friend, we are participating in the same sacred realm of language that has served us for most of the last forty thousand years.

---

5   Joseph Campbell, *The Flight of the Wild Gander: Explorations in the Mythological Dimension* (Novato: New World Library, 2002), 25.

6   Matthias R. Mehl, Simine Vazire, Nairán Ramírez-Esparza, Richard B. Slatcher, and James W. Pennebaker, "Are Women Really More Talkative Than Men?," *Science* 317, no. 5834 (July 2007): 82.

We have written records of human language as far back as five thousand years or so, but since we understand the development of writing is dependent on an existing spoken language, researchers have worked for decades to uncover the oldest evidence of spoken language. What makes the study of language so difficult is that the evidence is scarce. After all, there are no fossils of spoken language. What we do know is that something important occurred between one hundred thousand and forty thousand years ago. How do we know? Because we have cultural artifacts of art and ritual objects that lead us to believe civilizations were born.

DIGGING DEEPER INTO LINGUISTICS

Did we suddenly develop the capacity for language? Or were we slowly evolving all the while? Some researchers believe language developed in stages. American linguist Ray Jackendoff speculated that language developed over time in the following sequence:[7]

1. "In an early stage, sounds would have been used to name a wide range of objects and actions in the environment, and individuals would be able to invent new vocabulary items to talk about new things."

2. Next, individuals would "'digitize' signals into sequences of discrete speech sounds—consonants and vowels—rather than unstructured calls. This would require changes in the way the brain controls the vocal tract and possibly in the way the brain interprets auditory signals."

---

7   Ray Jackendoff, "How Did Language Begin?," Linguistic Society of America, n.d., https://www.linguisticsociety.org/content/how-did-language-begin.

3. "A next plausible step would be the ability to string together several such 'words' to create a message built out of the meanings of its parts. This is still not as complex as modern language. It could have a rudimentary 'me Tarzan, you Jane' character and still be a lot better than single-word utterances."

4. "A final change or series of changes would add to 'protolanguage,' a richer structure, encompassing such grammatical devices as plural markers, tense markers, relative clauses, and complement clauses ('Joe thinks that the earth is flat.'). Some hypothesize that this could have been a purely cultural development, and some think it required genetic changes in the brains of speakers."

Other researchers believe that language came in a single leap. A study by New Zealand biologist Quentin D. Atkinson, which applied mathematical methods to linguistic data from over five hundred languages around the world, suggested two key findings: Language originated only once, and the specific place of origin may be southwestern Africa.[8] Atkinson posited that if the African population began their migration to Asia and Europe from Africa sixty thousand years ago, the spoken language might also have existed. His theory suggests that not only is language perhaps older than we thought, but it might also have been the catalyst for our ancestors' dispersion and subsequent migration.

---

8   Quentin D. Atkinson, "Phonemic Diversity Supports a Serial Founder Effect Model of Language Expansion from Africa," *Science* 332, no. 6027 (April 15, 2011): 346–349.

No matter when or where language developed, we know that in its earliest form, language most likely used visual pictures. Some theorists argue that cave paintings found in Europe, Asia, and Africa ranging from 35,000 to 10,000 BC were the earliest attempts to communicate through visual symbols—a type of protolanguage.[9] Joseph Campbell visited such caves, particularly the "temple caves" of southern France and northern Spain. He speculated the cave images were created seventeen thousand to fifteen thousand years ago. Like my own experience in the vaults under the University of Aberdeen, Campbell acknowledged that, upon entering the caves, he left the light of the profane realm to enter the darkness of a more sacred one: "A terrific sense of claustrophobia, and simultaneously of release from every context of the world above, assails the mind impounded in those more than absolutely dark abysses, where darkness no longer is an absence of light but an experienced force."[10]

Figure 1.1. The ceiling of a "temple cave" in Lascaux in southern France, believed to have been created fifteen thousand to seventeen thousand years ago. Photograph courtesy of Joseph Campbell Foundation, https://www.jcf.org/mythblast-paleolithic-cave-art-time-and-eternity/.

Over time, humans combined the same visuals Campbell saw in the caves with storytelling as a way to understand the larger world and their

---

9   "When Did Human Language First Appear?," Study.com, January 2, 2018, https://study.com/academy/lesson/when-did-human-language-first-appear.html.

10   Joseph Campbell, *Primitive Mythology* (Penguin Books, 1991), 66.

own places within it. These stories, the same ones studied by Mead and Campbell, helped initiate the younger children about the traditions, rituals, customs, taboos, and values within the society. Eventually these stories were combined with written symbols, which meant that adult individuals could capture the stories they'd heard around the campfire as children. They could record them, share them, and build on them. Suddenly stories became timeless and universal. The wisdom of one generation could be harnessed and passed to the next generation. Cuneiform, the oldest known system of writing, dates back to around 5,300 BC and was developed by the Sumerians in ancient Mesopotamia. Though it looks unrecognizable, this ancient writing form would lead to the written language we use in the modern world today.

Figure 1.2. Some scientists believe cuneiform is the oldest known system of writing, was developed by the Sumerians, and dates back to around 5,300 BC. This relief is from Assyria and is exhibited in the British Museum, London. https://www.worldhistory.org/cuneiform/

The Power of Language

All things evolve. In fact, change is the one constant. The theory of evolution by natural selection, proposed by Charles Darwin and Alfred Russel Wallace in the nineteenth century, is a scientific theory that can be expanded to include language. In its simplest form, this "survival of the fittest" theory shows that when an environment changes, the traits that enhance survival in that environment must also gradually change or evolve. The same is true for all elements of the universe—even soil. As mathematical cosmologist Brian Swimme once noted, there is no soil on the moon, sun, planets, or anywhere else in the surrounding trillion miles. The earth itself required four billion years to create soil. In the same way, language has evolved and changed as human traits and cultural needs shifted.

Language is power, and according to growing research in linguistics, it has always held power. As we look at the evolution of communication, each time language evolved, so did humans. All human evolution has been reflected in its use of language. Each time we were able to use language to communicate in new ways, we held a new vision of what was possible. Each time we were able to use it in a novel, innovative way, we saw an explosion in the connections we could make between ourselves and the outside world.

Some scientists feel like it was the practice of language, along with organized hunting and gathering practices alongside the discovery of fire, that was a catalyst for humans' development. American anthropologist Clifford Geertz said, "By submitting himself to governance by symbolically mediated programs for producing artifacts, organizing social life, or expressing emotions, man determined, if unwittingly, the culminating stages of his own biological destiny. Quite literally, though quite inadvertently, he created himself."[11]

For Geertz, language not only helped create the advanced human species, but it was also essential to cultivating the human experience. Without it, he claimed humans would be "virtually ungovernable, a mere chaos of pointless acts and exploding emotions, his experience virtually shapeless. Culture ... is not just an ornament of human existence but ... an essential condition for it."[12]

If language helped advance our species as it changed from visual symbols, to storytelling, to written language, how might it advance us next? What will future linguists regard as the moment modern humans "inadvertently" created an evolved, superhuman version of ourselves? What artifacts will line the future vaults of the University of Aberdeen? The logical next chapter in the story of the evolution of language observed, recorded, and celebrated by the university since 1495 is natural language automation.

The Emergence of Natural Language Generative AI

The emergence of natural language generative AI was not a random event. The university has spent the last 520 years writing its chapters in the timeless story of the evolution of language. From the two hundred thousand priceless literary treasures preserved in the vaults emerged the revolutionary technology we now call Arria. But before we can understand the future of natural language technology, we must first understand its past.

---

11 Clifford Geertz, *The Interpretation of Cultures* (Basic Books, 1973), 53–4.

12 Geertz, *The Interpretation of Cultures*, 51.

In 1495, Pope Alexander VI penned a papal bull with the intention to found a university—not a Catholic one, surprisingly—but one *that would be open to all and dedicated to the pursuit of truth in the service of others.* He entrusted this decree to Bishop Elphinstone, a man dedicated to the power of language. Elphinstone, along with his father, believed that, through language, primitive humans could become noble and God fearing. It was this belief that led the father and son all over Mesopotamia collecting two hundred ancient artifacts of language—from diaries to scrolls to gold-leaf drawings. When Elphinstone was entrusted with the papal bull, he made an epic journey from Rome to Aberdeen in Scotland. At the same time as his stonemasons were laying the foundations of his new seat of learning, he bequeathed to the university his collection of artifacts, which included his own manuscripts. It was his hope that other unwavering believers like himself might hold the vision for a more evolved language and thus a more noble human race.

Bishop Elphinstone laid the original foundations of the university's enduring dedication to the evolution of language. Consistent with his early intentions, over the past five hundred years, the university has amassed an extraordinary collection of artifacts, which has grown from two hundred to two hundred thousand priceless literary treasures, that trace the evolution of written language back as early as the third century BC. These treasures, together with Bishop Elphinstone's manuscripts and the original papal bull, are all preserved in the special vaults located under the library.

Figure 1.3. William Elphinstone was a Scottish bishop and statesman, and the founder of the University of Aberdeen. While William Elphinstone's main contributions were in statesmanship, diplomacy, law, and education, he played a crucial role in introducing the printing press to Scotland in 1507. This introduction to printing significantly impacted the spread of knowledge and potentially influenced the evolution of language by making texts more accessible and promoting literacy.

We are fortunate at Arria to share Bishop Elphinstone's vision with the University of Aberdeen. Their historical campus serves as our Center of Excellence, housing our development team who works on a future technology that has been with us since the past—converting data into language. From Aberdeen's ancient school of learning has arisen one of the most advanced technologies ever developed.

The scientific foundation for Arria's technology is based on more than thirty years of research and development by our founding scientists. With two hundred thousand artifacts preserved in the dark vaults below, our language scientists' work—grounded in the past—holds a vision for the future of equitable access of information that empowers rather than overwhelms us, all in pursuit of truth in service to others.

We've been on this planet for two million years, and in that time we have constantly reinvented ourselves. Whether through agriculture, industry, language, or technology, we have advanced what it means to be human. During that time, however, some things have remained the same as it was for our ancestors sharing stories by firelight: We pursue connection with our culture and the people around us, we seek an understanding of our place in the larger world, and we deem to share that with the next generation. It is instinctual to use language, our greatest power, to share our stories so they might be remembered and added to, creating a larger story of humanity.

As mathematical cosmologist Brian Swimme states, "'Story' is the fundamental answer we give when we are asked what really matters in this world."[13] Even surrounded by information as we are today, our greatest contributions remain connection, communication, and evolution. For this reason, we must continue to hold a vision of something larger for which to strive. Without this vision, this purpose, this expansion, we aren't growing; we're merely existing. In the beginning there was the word, and as our Arria technology evolves, we will remain dedicated to keeping language as the cornerstone of humanity, to building on the legacy of human stories, and to pursuing truth in service to others.

---

13 Brian Swimme, "The Resurgence of Cosmic Storytellers," *NAMTA Journal* 23, no. 1 (Winter 1998): 145–56.

CHAPTER 2

The Evolution of Information Delivery

When I was a young child, we owned a technological wonder: a black-and-white TV that was almost as big as I was. The wooden television box loomed large in our living room, and we marveled at how easy it was to walk up and turn the set's dial to change channels. We couldn't believe our luck to have *four* channels to choose from—what a revolutionary invention! During this time, there were other analog innovations that made our lives better, like our toaster oven, microwave, and diamond-needle turntable. Over time, we upgraded to an analog eight-track tape and cassette player. What do these bulky analog artifacts of the past have to do with the future? They remind us that the only constant is change.

As we introduced earlier, the law of perpetual motion tells us that all things are forever changing, even technology—and specifically *information* technology. Humans' role has been to enjoy the benefits of these evolving technologies and to reinvent our own lives each time a new innovation unburdens us from a routine behavior and expands life experiences to a whole new level. When we could move the power of cooking for hours to a microwave, for example, our time was freed up to do other things for ourselves and our families. When my music collection transitioned from a proud cabinet of vinyl albums containing my favorite artists neatly categorized A–Z, to a cloud-based subscription providing unlimited access to anything I could think of listening to—not to mention suggested music based on trending favorites—it moved the experience of entertainment beyond what I would have ever imagined. The entire music industry was disrupted, digitized, dematerialized, and, ultimately, democratized. Interestingly enough, turntables and albums are back! This continual transformation has allowed humans to learn from the technologies of the past, enjoy their conveniences today, and anticipate the new experiences and the advances of tomorrow.

The Sequence

Information delivery essentially sits on top of big data analytics and makes sense of it all. The automation of information delivery is in its third and most important phase, though its evolution began when we moved the power of numbers to the machine and created tabulation. In the professional realm, we were freed from doing complex mathematical equations and could spend time with customers or on other projects. In the personal realm, we spent less time at the kitchen table doing stack additions and subtractions while making budgets and paying bills. Spreadsheets, however, cannot explain themselves. They require further analysis and explanation. Even so, this small reallocation of one power—the power of numbers—became the template for future advancements in information delivery.

THE POWER OF NUMBERS	THE POWER OF PICTURES	THE POWER OF ACTIONS	THE POWER OF LANGUAGE	THE POWER OF VOICE
Data Analytics 1.0	Data Analytics 2.0	Data Analytics 3.0	Data Analytics 4.0	Data Analytics 5.0
TABULATION	VISUALIZATION	AUTOMATION	NARRATION	CONVERSATION

Figure 2.1. The evolution of information delivery

The next power we allocated to machines was the power of pictures. Early visualizations—like cave drawings—were profound breakthroughs in the establishment of language. Though these etchings seem rudimentary to us, they were innovative marvels at the time and reflected humans' growing reverence for the power of language. In fact, without this small step, language might never have evolved into the sophisticated tool that defines humanity, and humans might not have evolved past the early cavemen. In this way, words were the true fire.

When machines assumed the power to create visualizations, we could represent data through graphs and charts, which allowed us to pull value from a single source rather than poring over several at once. Visualizations, however, also required further analysis and explanation, usually by experts.

With visualizations enhancing humans' productivity, we were able to move the power of actions to the machine, which paved the way for automation. There is no doubt this evolutionary step freed up our time and mental energies to do more and be more and allowed us to transition the power of language to the machine, which led to the birth of narration as an innovative form of information delivery. As we will learn, though business visualizations—turning numbers into visuals—remain one of the greatest advancements in the computer era, turning numbers into language is even more of an astonishing accomplishment.

Arria starts where visualizations stop and makes data automatically communicate its meaning. As history shows, the output of the most advanced analytics in the world will one day soon be seen as the cave drawings of the digital age.

How Did Natural Language Automation Emerge?

The evolution of language is much like the evolution of any art form. Though it may have evolved over centuries by the efforts of countless humans, for the artist herself the only canvas that matters is the one before her. Up close, she adds one dab after another to the image. She is not creating art, much less adding to the evolution of the form; she is merely applying colors onto a blank page, one singular stroke at a time. When she pulls back, however, those bits create a complete portrait and add to the ever-unfolding evolution of art. In the same way, when Arria's founding scientists at the University of Aberdeen are working, it feels like they are adding single dabs to a canvas. It's only with a zoomed-out perspective that they understand they are exteriorizing the timeless power of language and evolving how information is delivered.

Many believe that Arria's natural language is a new technology born from a random event. Once we consider the evolution of language, however, it's clear that natural language generative AI was the logical next chapter. The modern-day foundation for the emergence of language technology took form in 1972 when, at the dawn of the digital revolution, University of Aberdeen leaders identified computing as a key field of scientific endeavor. What we now refer to as the science of natural language generation (NLG) truly took shape in 1995, when Ehud Reiter founded the Aberdeen NLG Group. This game-changing group grew to become the largest in the world in terms of people, publications, and projects. One outcome of the group's work was defining the goal of the science of language automation: *to discover how to give machines the ability to receive inputs, reason over those inputs, determine what is important, and communicate the insights to others using natural language.*

Considering this evolution of language, how was it that scientists like Ehud Reiter understood as far back as 1995 that NLG was the obvious next step in the evolution of language? Because they understood that NLG was an innovation but also a prerequisite for the next advancement.

THE EVOLUTION OF LANGUAGE

1495   Bishop Elphinstone makes an epic journey from Rome to Aberdeen in Scotland.

1972   University of Aberdeen leaders identify computing as a key field of scientific endeavor.

1995   Arria founding scientist, Professor Ehud Reiter, founds the Aberdeen NLG Group.

2000   Reiter coauthors and publishes *Building Natural Language Generation Systems*, which goes on to be widely considered as the seminal textbook on NLG. Also in 2000, Arria founding scientist Dr. Yaji Sripada joins the University of Aberdeen NLG Group.

2009   The university forms Data2Text Limited to realize the global commercial potential of the breakthrough NLG software technologies flowing from its computer science labs and world-leading NLG scientists.

2013   Data2Text Limited is acquired by Arria. Both the University of Aberdeen and Arria's founding scientists are significant founding shareholders in Arria.

As the Aberdeen Group began making breakthroughs, the rise of data began gathering its momentum and, as history now shows, set the stage for not only the emergence of NLG as a breakthrough AI technology but also the demand of the convergence of natural language technologies, and all the other interdependent technologies that power the datasphere as we know it today. The Aberdeen Group recognized that having data all around us does not benefit us unless we can unlock its insights through the power of aggregation, correlation, comprehension, and, ultimately, communication.

In 2009, the university, together with Arria's founding scientists and others, formed Data2Text Limited to realize the global commercial potential of the breakthrough language technology flowing from its computer science labs and its world-leading scientists. Other tech giants were poised to buy Data2Text, but the university wanted a group whose mission was to take the power of language everywhere—a group of unwavering believers. They didn't want it to merely be owned; they wanted it to be shared for the global and evolutionary impact it could have on humans. In 2013 Arria completed the acquisition of Data2Text and its intellectual property.

Language Accelerates Understanding

What data scientists have long understood is that comprehension requires two steps: access and communication. If data is outside of our area of expertise, for example, it's essentially locked. Its insights get unlocked, however, when we can (a) access the information, and (b) communicate the insights across a family, community, organization, or culture.

The best way to understand new concepts is to have it presented through language to accelerate understanding. When we read it, we understand it. Without language, it is impossible to get real value from data, much less share it with others. With language, we can be certain that we extract insights and create a shared understanding with others. Language allows data to become a universal form of information delivery. It's universal because we don't have to read programming language, interpret a graph, or digest a spreadsheet. Arria's natural language technology allows us a universal communication platform for understanding that can be delivered with others, no matter our skill sets or native language.

Data scientists create impressive dashboards of analytics, but until you add language to share the insights, it's incomplete and open to interpretation. It's only valuable to a small number of specialized people. We would glaze over the data, even if it was directly applicable to our lives, because we couldn't identify the relevant aspects, much less extract meaning from it. Arria is the bridge between data and insights, between humans and understanding. Without the bridge, you cannot access the other side.

Arria can identify which parts of data analysis are most useful for a particular audience and can customize the relaying of the information accordingly. The insights are communicated in conversational language for the specific audience using all relevant information. For example, in 2011, the Aberdeen NLG Group worked to turn data from an electronic patient system into reports for the Royal Infirmary of Edinburgh's neonatal intensive care unit. As we will explore fully in a later chapter, understanding the need for different reports for each audience, the group created three reports: one for doctors and nurses to assist in decision-making, one for nurses to offer one another at shift changes, and one for parents that included information relevant to the care of their child.

What's Next?

The capabilities of NLTs are far reaching for many industries, and the full potential of NLT is still being explored. What we do know from the evolution of language and information delivery is that each time humans moved one power to the machine, it proved to be a prerequisite to the adoption of the next innovation. In this way, the creation of one technology enabled the creation of the next. Knowing what we do about the ever-changing world, we understand that the output of most advanced analytics in the world will soon be considered the cave drawings of the digital age.

Let's consider the multi-billion-dollar category of business intelligence (BI) dashboards, which are used to track metrics and manage key details. Though these were indeed an innovative marvel for the business world at one time, we are now in the augmented era of analytics. Visualizations alone only do half of the work because they still require interpretation. Without analysis and interpretation, all the metrics in the world are useless. No matter how long business leaders stand before a BI dashboard, they are still reliant on analysts to uncover and communicate insights. It's the combination of the dashboard with NLTs that unlocks the data and reveals its value through data storytelling. We're moving out of the cave and into the future, helping data evolve through the power of language—humanizing the way we interact with data.

Data storytelling provides the ability to combine interactive data visualization with narrative techniques in order to package and deliver insights in a compelling, easily understood form for presentation to decision makers. The enhanced AI-driven experience includes all the NLTs, which fall under the umbrella of natural language processing (NLP). Natural language query (NLQ) enables users to ask questions of the data using terms that are either typed into a search box or spoken, natural language understanding (NLU) reveals the sentiment of the data and therefore the emotion or tone, and of course NLG makes it possible to automatically create linguistically rich descriptions of insights found in data. Within the analytics context, as the user interacts with data, the narrative changes dynamically to explain key findings or the meaning of charts or dashboards. NLT, machine learning, and hyperpersonalization are part of the underpinnings of true, advanced generative AI and agentic AI systems. Converging these technologies creates a powerful combination, automatically turning data into written narratives that clearly explain insights.

Where there's data, there is the need to extrapolate meaning from it. Since we have moved the power of voice to the machine through the advancements of text-to-speech, we are moving toward the newest information advancement: conversation. Imagine if your data could not only speak but listen and converse naturally. If you're an executive, for example, rather than tracking down your human analyst for insights, what if you could ask a digital assistant the same questions and get an up-to-minute briefing, specifically catered to your needs, about your sales results or pertinent metrics? What if you could get responses to your follow-up questions that analyze all relevant data and even offer further information that the digital assistant learns is of interest to you? Within minutes, you are fully informed about which branch offices or sales managers require your executive support and attention. Now, imagine how this same technology could benefit your suppliers and employees. As we will explore in part II, these technologies benefit your organization's productivity, efficiency, and efficacy.

The most forward-looking organizations are considering technology as a better way to serve up insights to decision makers. Arria's technology analyzes data and communicates insights to humans in a natural way, turning data into narratives, explanations, summaries, and conversational content that is indistinguishable from human writing at machine speed and scale. Arria's primary use cases include report automation and the generation of narrative insight summaries and explanations. These uses allow organizations to eliminate manual processes, scale efficiently, accelerate decision-making, democratize data understanding, and empower unified planning and decision-making efforts.

Despite its humanlike language qualities, the technology, in fact, uses a machine, so humans reap the benefits of having extraordinary amounts of data analyzed and presented in the best way possible with high levels of accuracy. This frees humans from highly repetitive and tedious work.

In fact, research found that Arria automates 80 percent of manual reporting, which equates to a time saving of up to 60 percent for analysts and decision makers.[14] This is precisely why AI technologies have the ability to empower humans: Hu + AI = superhuman.

Language has been uniquely human and, in fact, defines humanity. From the two hundred thousand priceless literary treasures preserved in the University of Aberdeen's vaults emerged the revolutionary technology we now call natural language generative AI. At Arria, we understand the awesome responsibility we have in evolving language from the exclusive territory of the human mind to that of the digital mind. Today, both the university and Arria's founding scientists are significant founding shareholders in Arria, and we remain committed to the shared mission of the university to keep language technology

---

14 "The Total Economic Impact of Arria Natural Language Generation Study," Forrester Research, 2021, https://go.arria.com/forrester-tei-report.

open to all and dedicated to the pursuit of truth in the service of others. By respecting the cultural ethos, beliefs, and history of the founders of the University of Aberdeen and their dedication to language, we enable great institutions to remain in existence through generations, during which time everything around them changes in infinite ways. Building an enduring global commercial enterprise that will honor the scientific legacy that gave rise to our mission is the deepest intention of the Arria team.

The human species has been evolving since the beginning of our existence, and language has always been a part of that evolution. Part of the universe's purpose—and ours by extension—is to change and evolve. As we will explore in the next chapter, there is always something higher to reach, something bigger to work toward. The underlying universal principle of forward motion reminds us that growth, hope, movement, evolution, and innovation are our natural, innate drives. As John F. Kennedy said in a 1960 speech, "Efforts and courage are not enough without purpose and direction."[15] Without forward motion, you're static. From the beginning of time to the beginning of your day, life will present challenges, and each one propels you to your own individual evolution. Arria has the capacity to enhance that evolution by bridging the gaps toward understanding, thereby aiding our evolution as a species. If language was one of the greatest advancements that catapulted humanity, then bringing language to technology surely will have a similar impact.

With the advent of natural language technologies, for the first time in human history, the power of language, the very thing that defines our

---

15   Papers of John F. Kennedy, Pre-Presidential Papers, Senate Files, Series 12, Speeches and the Press, Box 911, September 17, 1960, Coliseum, Raleigh, North Carolina, https://www.jfklibrary.org/asset-viewer/archives/jfksen-0911-055.

humanity, has now been captured in software, moving outside of the human mind. Nothing will ever be the same again. As the power of language moves rapidly from the mind of man to the machine in the form of software, life will change for everyone, everywhere, in ways that cannot yet be imagined.

Natural Language Technologies: A Renaissance and Paradigm Shift

The rise of NLTs marks a paradigm shift in human-computer interaction. While the internet revolutionized access to information, NLTs are transforming how we understand, reason, and interact with knowledge itself. Many experts now consider this development a breakthrough of greater significance than the internet, a true renaissance of human-machine communication.

Natural language technologies have realized the long-held vision of allowing machines to understand and communicate in human language. This shift has reinvigorated the way people interact with technology, enabling more intuitive, intelligent, and accessible systems across every domain. The transformation manifests in four fundamental ways: Language has become the new interface, replacing complex commands with natural conversation; ubiquitous intelligence now allows AI capabilities to be accessed anywhere through simple speech or text; creative collaboration between humans and machines opens new possibilities for innovation; and cognitive amplification extends our mental capabilities beyond their natural limits.

This progress is a cultural and cognitive awakening—a paradigm shift from information to understanding.

The internet connected people to vast troves of information but lacked true understanding. NLTs bring about a new era where machines interpret context, infer intent, and engage in meaningful dialogue through four key capabilities: intent recognition that allows systems to understand what users truly want, adaptive learning that enables continuous improvement from interactions, knowledge synthesis that combines information from multiple sources into coherent insights, and agentive AI that can take autonomous action based on understanding.

With NLTs, we are shifting from searching for information to conversing with understanding. Where the internet democratized access to content, NLT democratizes cognition and decision-making.

INTERNET REVOLUTION VERSUS NATURAL LANGUAGE TECH REVOLUTION	
Access to information	Understanding of information
Human adapts to systems	Systems adapt to humans
Static content	Dynamic, generative interaction
Information overload	Contextual synthesis

The internet was about communication; natural language technologies are about comprehension and action. The convergence of natural language technologies within the umbrella of generative AI and agentic AI has immediate and profound implications across multiple domains. These systems enable the automation of cognitive tasks, freeing humans for higher-level strategic thinking while ensuring the accuracy of complex processes like regulatory report automation. They deliver operational efficiencies that accelerate actionable intelligence organizations can trust, while simultaneously creating personalized AI tutors that democratize access to subject matter expertise. Perhaps most significantly, they automate the delivery of personalized information at scale, presented in natural language that adapts to each individual's needs and context.

The Age of Understanding

Like the printing press and the internet, natural language technologies are ushering in a new era—the Age of Understanding. Machines today go beyond simply learning human language to actually understanding humanity. This transformation empowers individuals and organizations to access, generate, and act on knowledge in real time.

NLTs are poised to transform how enterprises and government agencies operate by streamlining processes, enhancing decision-making, and enabling smarter services. The result is a shift from reactive to proactive governance and from siloed workflows to fully integrated systems. As NLTs become embedded in high-stakes domains such as enterprise, healthcare, law, finance, and public policy, the importance of accuracy becomes paramount and is foundational to responsible AI deployment. Misunderstandings, hallucinations, or ambiguous outputs can have serious consequences. Ensuring precision, transparency, auditability, and control is essential for building public trust and achieving real-world impact.

As we pointed out at the start of the chapter, having a system that can keep up with rapid change can make or break an organization. Arria has the flexibility to integrate the strengths of any LLM's machine learning language model with our own natural language technology and methodology. The goal is mainstream adoption of the ability to transfer knowledge in words by leveraging the best features of the language technologies available. In combining Arria's deterministic approach and LLM's probabilistic approach, we allow enterprise companies to leverage the strengths of both generative AI technologies to query and understand data in a controlled, deterministic way, combined with the power to search and retrieve internal documents and external information chosen and approved by the user (human-in-the-loop) for summary by the LLM.

Arria's mission and purpose is to be the bridge between humans and the information and understanding they seek by providing data-driven, responsible AI solutions. The pace at which generative AI is advancing is extraordinarily fast—arguably faster than almost any other major technology in recent history.

NOTES FROM AN UNWAVERING BELIEVER

Lyndsee Manna, generative AI executive and thought leader

Although all species have their ways of communicating, humans are the only ones who have mastered cognitive language communication. Language allows us to communicate what we know and transfer that knowledge and sentiment to others.

Every day we communicate and consume language with such apparent ease that we take it for granted. Starting around the age of eighteen months, most children begin communicating in rudimentary sentences like "More water," or "Hands sticky." Many linguists agree with Noam Chomsky's assertion that language is innate and that children are prewired with a universal grammar.

Though they might unconsciously assimilate grammar rules from the people around them, children learn new words and patterns through interaction with other humans. They master language through pattern recognition and rapid learning, thanks to the adults and other humans around them. This modeling trains their brains how to use language appropriately (these are known as language models).

The human brain learns how to calculate when and how to use language and continues to evolve its own language models constantly over time through training. Natural language technology emulates how a human calculates when to use language—what we call language analytics—and how to use and put the words together—what we call cognitive linguistics. This is exactly how Arria technology works. We emulate the way a human brain calculates what to say, when, and why.

How we express our knowledge in language is fluid and constantly changing as what we know (data) changes. Although we can dynamically change the way in which we communicate to adapt to different constituents with varying levels of comprehension, humans can only do one task at a time. Through the use of Arria technology, humans' ability to communicate and transfer knowledge dynamically can be done at machine scale and speed, processing large amounts of data and generating individualized communication in language that is dynamic and unique. This automation enables knowledge to be personalized and shared exponentially in ways that each individual can comprehend.

CHAPTER 3

WordCode: Language as an Evolutionary Tool

Several years ago, I was traveling with a business partner after an off-site presentation. We were flying over the Midwest when my partner said, "Great job today, Sharon, but you tend to get too warm and fuzzy during your presentations at times."

Having always believed that being yourself is the easiest way to be, I admittedly lead with passion and enthusiasm. "I think it's important to connect and build relationships, not just talk about the business side," I answered.

No sooner had I given him my perspective than the flight attendant announced, "I hope everybody's having a great day today. Why don't you take a moment and just share some love with your neighbor?"

I turned to my partner and said, "Warm and fuzzy seems okay for this airline!"

Words have power, and how we deliver words also has power. A parent who hears their child say "Mama" or "Dada" for the first time understands the power of words. A patient who articulates a shrouded fear to their therapist for the first time understands the power of words. A person who hears their beloved whisper "I love you" for the first time understands the power of words. And conversely, a child who hears an adult spew "You're stupid" also understands the power of words. Words have the capacity to heal us or hurt us. In fact, the memory of some words can outlive the humans who spoke them.

The words we choose shape our world. Every word that crosses our lips has power. This is not merely the opinion from someone who admittedly loves words; it is, in fact, a universal law. To understand the power of words, we must first understand the law of vibration, which states that everything in the universe is vibrating at a subatomic level. Simply put, we are what we are vibrating. If we consider that even particles are vibrating, then it makes sense that high-vibrating particles are attracted to other high-vibrating particles. This is the scientific principle behind "like attracts like" and "your vibe attracts your tribe." The law of vibration then is directly connected to the law of attraction. These laws work without fail because that's what laws of the universe do. Like gravity, we don't have to see it or believe in it for it to be real. The law is in action whether or not we're on board.

Using these fundamental principles, it's easier to understand that what we put out into the universe also dictates what we draw to ourselves. As a CEO, if I lead with positivity, I will contribute to a thriving environment that draws those same qualities back to me. The words we choose, then, are a further extension of these universal principles and contribute to the vibrational energies we are putting out into the universe and can direct the trajectory of our collective evolution.

In the book *The Hidden Messages in Water*, Japanese scientist Dr. Masaru Emoto explores how words have vibrational effects on the molecular structure of water. When a word was spoken—even written or thought about—in the direction of water, Emoto noted and photographed changes in its crystal structure. Japanese words for "thank you" and "love" and "gratitude" created lovely symmetrical hexagons. On the contrary, words like "you fool" or "war" produced misshapen crystals. He also observed how water crystals altered their structure after a prayer or blessing was offered to the water.

From his research, Dr. Emoto concluded that if words like "love" and "gratitude" could alter the water crystals in such profound ways, how might he alter his own environment with positive words and thoughts? Since that time, Dr. Emoto refers to individuals who shine with light as Aikansha-bito ("beautiful humans of love and gratitude"). As an unwavering believer himself, he posited that if all humans embraced this same perspective and consciously chose their words and the emotions behind them, we might be able to make the entire world more loving and gracious.

Figure 3.1. Water crystals exposed to the words "I can't" (left) and "I can" (right).
Images courtesy of Office Masaru Emoto.

The reality is no matter how much we take language for granted, the vibrations of words have the power to alter us on subatomic levels. In fact, after studying the vibrational impacts of sounds, New Zealand researcher Meera Raghu concluded, "Whether we like it or not, the vibrations of these sounds reach us, not only through the hearing sense but also by coming into contact with the physical body. The sound vibrations can affect us either positively or negatively, entering into our being, via the physical, mental, and emotional realms, thereby affecting our consciousness as a whole."[16] If words have the power to change water crystals and subatomic particles, then imagine the ramifications this can have on an individual's life, or on humanity as a whole. Language is not just a series of clicks and vocalizations; it's a powerful tool for the evolution of the human species.

DIGGING DEEPER INTO RESONANCE THEORY

Scientists understand that all things in our universe are constantly vibrating, even those that appear stationary. So what makes them attract other particles with the same vibration? Over the last decade, new understanding about resonance theory of consciousness explains the interesting phenomenon that occurs when different vibrating particles come into proximity—when they essentially "sync up" and vibrate together at the same frequency. This is the mystery that is called spontaneous self-organization.

---

16   Meera Raghu, "A Study to Explore the Effects of Sound Vibrations on Consciousness," *International Journal of Social Work and Human Services Practice* 6, no. 3 (July 2018), http://www.hrpub.org/download/20180730/IJRH2-19290514.pdf.

In his book *Sync*, professor and researcher Steven Strogatz provides examples of what he calls "sync" or synchrony, including the following:

- Certain species of fireflies, when gathered together in large numbers, flash their lights in sync.
- Laser beams are the result of trillions of atoms emitting in concert.
- The tides have locked the moon's orbit so that its axis turns at the same rate in which it circles the sun—the reason why we always see the moon's "face" instead of its dark side.
- Human brains have large-scale neuron firings at certain frequencies, resulting in rhythmic pulses associated with seizures.

Strogatz says that even though we don't yet understand "the tendency to synchronize," it remains one of the most prevalent impulses in the universe, encompassing atoms, animals, people, and planets.[17]

On the surface, these phenomena might seem unrelated. After all, the forces that synchronize brain cells have nothing to do with those in a laser. But at a deeper level, there is a connection, one that transcends the details of any particular mechanism. That connection is mathematics. All the examples are variations on the same mathematical theme: self-organization, the spontaneous emergence of order out of chaos. By studying simple models of fireflies and other self-organizing systems, scientists are beginning to unlock the secrets of this dazzling kind of order in the universe.

---

17 Steven Strogatz, *Sync: How Order Emerges from Chaos in the Universe, Nature, and Daily Life* (Hyperion, 2003), 14.

Whether we understand these resonating principles or not, they are at work in our lives every day and perhaps play a larger role in the realities that surround us than we may ever know. If we don't consciously use these principles, we are missing opportunities for growth and evolution.

The Power of Understanding

At the age of nineteen months, Helen Keller lost her sight and hearing. For five years, she communicated as best she could, but living without language made her feel like "a phantom living in a world that was no-world."[18] When Keller was seven years old, her new teacher, Anne Sullivan, arrived. Sullivan was not only a teacher; for she, too, felt like a "phantom" when a childhood illness impaired her vision at five years old. It was Sullivan who was able to bring Keller from the "no-world" to the known world through the power of language. Of Sullivan's arrival, Keller wrote:[19]

> I was like an unconscious clod of earth. There was nothing in me except the instinct to eat and drink and sleep. My days were a blank without past, present, or future, without hope or anticipation, without interest or joy. Then suddenly I knew not how or where or when, my brain felt the impact of another mind, and I awoke to language, to knowledge, to love, to the usual concepts of nature, good, and evil. I was actually lifted from nothingness to human life.

Before language, Keller felt like a primitive creature driven totally by her senses. The catalyst for her awakening to "human life" was one

---

18   Helen Keller, *Teacher: Anne Sullivan Macy: A Tribute by the Foster-Child of Her Mind* (Greenwood Press, 1985), 8.

19   Helen Keller, *Light in My Darkness* (Swedenborg Foundation, 2000), 5–6.

simple word. As the story goes, Keller was confusing the word "cup" with "water" when Sullivan took her to the water pump. She plunged one of Keller's hands under the rushing water, and on the other hand she traced *w-a-t-e-r.* According to Keller:[20]

> She spelled **w-a-t-e-r** emphatically. I stood still, my whole body's attention fixed on the motions of her fingers as the cool stream flowed over my hand. All at once there was a strange stir within me—a misty consciousness, a sense of something remembered. It was as if I had come back to life after being dead! I understood that what my teacher was doing with her fingers meant that the cold something that was rushing over my hand was water, and that it was possible for me to communicate with other people by these hand signs... Now I see it was my mental awakening. I think it was an experience somewhat in the nature of a revelation... That first revelation was worth all those years I had spent in dark soundless imprisonment. That word "water" dropped into my mind like the sun in a frozen winter world. The world to which I awoke was still mysterious; but there were hope and love and God in it, and nothing else mattered. Is it not possible that our entrance into heaven may be like this experience of mine?

For Helen Keller, the word that unlocked her potential was simple but profound: "water." What we all must understand, and as Keller so eloquently wrote in her later years, is that language is an inescapable part of humanity. Without it, we are locked in a "phantom world." Even if we aren't speaking aloud, our thoughts pass through the prism of language.

Whenever you look at a piece of art, for example, your interior thoughts filter through preconditioned language routines affected by your culture. In fact, if you think of yourself in software terms, your culture is your operating system. Everything else builds on top of

20 Keller, *Light in My Darkness*, 6–7.

that—from your language to your values to your traditions. As we know, any software sequence must be started from the outside. The same can be said of human evolution. In order to evolve, to find the "sun in a frozen winter world" as Keller writes, we need the benefit of an outside force to start the sequence. For many people, that external force is language.

As Keller learned that day by the water pump, the power of language can initiate a human's growth cycle. None of us might even know who Helen Keller is without that one word, dropped into her palm like a gift—"water." In the same way, we can all be "powered on" by a word or series of words. At Arria, we think of this as WordCode.[21] We all have a word or a series of words that motivates us and propels us toward our own evolution—our WordCode. If we have no WordCode acting as our vibrational orientation device, then we, too, might feel like "phantoms" in a "no-world." With it, however, we know our purpose and our power as humans, and we can use language's vibrational power to attract the life we want to live.

It makes sense that the power of language—which is the defining feature of human beings—would also be the thing to propel us toward something greater. Furthermore, when we apply the universal principle of the law of perpetual transmutation of energy—which states that even the smallest action can have a profound effect—then we begin to understand how words matter and *how much* they matter.

Having innovative language technology doesn't eclipse our humanity—it enhances it. For all the things NLTs can do and all the ways they can enhance our lives, they do not replace our own humanity. They can give us models of how to "reboot" ourselves, but

---

21   Thanks to my dear friend and poet Gerald Henry for creating and defining this term.

they are always a replica of that which cannot be replicated—love, connection, humanity.

Part of our essential humanity is the desire to connect through language. When someone is hurting, we offer words of kindness, we say prayers, we write letters, we share our own struggles with them. Though language is often taken for granted, it is truly the greatest gift we can offer to others. Many times in my life, I wondered what the point was of offering words to someone who wouldn't listen or who wasn't ready to listen. The answer is simple: Because sometimes the words we rattle off without thinking provide the listener all that they needed—their WordCode. It's not always the words themselves that help but the meaning and vibrational energies behind them. What if Anne Sullivan hadn't continued to offer words to Helen Keller? What if she withheld her words and their subsequent power? What if Keller remained a "phantom" in a "no-world"? How many of us would have evolved more slowly, or never at all, if not for the kindness of others to drop a word or series of words into our lives like a gift?

In Norman Maclean's story *A River Runs Through It* (made popular by the 1992 movie of the same name), there is a powerful moment at the end of the tale when the father is offering the eulogy at his son's funeral. His words remind us that we should never stop trying to offer our words—our connection to humanity and one another—to those we love:[22]

> Each one of us here today will at one time in our lives look upon a loved one who is in need and ask the same question: We are willing to help, Lord, but what, if anything, is needed? For it is true we can seldom help those closest to us. Either we don't know

---

22 Norman Maclean, *A River Runs Through It and Other Stories* (The University of Chicago Press, 1976).

what part of ourselves to give or, more often than not, the part
we have to give is not wanted. And so it is those we live with and
should know who elude us. But we can still love them—we can
love completely without complete understanding.

The only thing the father could offer his troubled son was his own
flawed, mortal, human self. And though we think sometimes we
shouldn't speak unless we have something profound to say, sometimes
it's the utterances themselves, the primordial human vocalizations,
that bring others peace through human connection. They are one
human alongside another human alive on this vast planet, whirling
through the galaxy; perhaps they are confused, perhaps they are lost,
but they are not alone. That's the power of words.

There's a saying that if everyone did what they could, it would be
enough. What if we all agreed to hold the same WordCode: *love*. How
might humanity evolve? If we approach a wayward loved one, we don't
need to understand anything other than the fact that they, too, are
human. This is why we have to say what needs to be said. We have to
offer our own fears, anxieties, and experiences. It will fall wherever it
will fall, and it will either benefit them or not, but leading with your
own WordCode of love will never be wrong.

Knowing our WordCode means spending less time worrying about
what to say because we are assured that we have the right intention
behind our words. Too often we get caught up in choosing the *right*
words rather than the *right kind* of words—the kind that are delivered
with love and compassion, the kind that create a more positive, aligned,
synchronized path toward our own individual and collective evolution.

I should also mention that sometimes our WordCode changes. In the
same way an operating system might need to be rebooted, so might

a person. There is no shame in this. In fact, the more WordCodes we resonate with throughout our lives, the more evidence we have of our own growth cycle—our own personal evolution. The good thing about adopting *love* as your WordCode is that you never outgrow it; in fact, it helps you evolve into your fullest potential.

The Heart of Technology

In a constantly data-driven, ever-evolving world, the greatest gift we can offer others is our own humanity. It's not about the words we choose; it's that we choose to offer words at all. It's that we take on the Herculean task of being present with another human. It's that we offer the greatest tool of connection humanity has ever known: language. It's not about what the words mean; it's about what they represent and the vibrational energies they release into the universe.

What is all this leading to at the end of our lives? Love, of course. As much as I'm an unwavering believer in the technological innovations of the future, love is impossible to replicate. We can reproduce the language generation technologies, but we can't imitate the human emotions behind the words. AI can duplicate the human mind but not the human heart. It's not just the words that have power; it's the essence of humanity behind them. If words can impact the structure of an ice crystal, what can they do for ourselves, our families, our organizations, and our communities?

Building a better, wealthier, more conscious, more cultured future for humanity means placing the most enduring, universal, cultural concept of all—love—at the heart of technology. This is why the book's title encompasses both the words "evolve" and "love"—for you cannot spell one without the components of the other. There can be no human evolution without love at its center. Technology's role in our lives is to aid this evolution.

Comparative mythologist Joseph Campbell says in *The Hero with a Thousand Faces*, "Religions, philosophies, arts, the social forms of primitive and historic man, prime discoveries in science and technology, the very dreams that blister sleep, boil up from the basic, magic ring of myth."[23] Technology doesn't fracture our humanity; it mends it. It boils up from some magic place where we are imperfect and fallible but also interconnected and whole.

Helen Keller also seemed to grasp how language was the bridge to a more evolved humanity when she wrote later in life, "The words 'love' and 'wisdom' seemed to caress my fingers from paragraph to paragraph and these two words released in me new forces to stimulate my somewhat indolent nature and urge me forward evermore."[24] My quest is to evolve no matter what I'm doing—that is what urges me forward. That's why my WordCode is *evolve*. Whatever your chosen word is, take responsibility for it. Own it. And then deliver that word—with love and gratitude—to the world.

---

23   Joseph Campbell, *The Hero with a Thousand Faces* (New York: Bollingen Foundation, 1949), 3.

24   Helen Keller, *How I Would Help the World* (Swedenborg Foundation, 2011), 29-31.

PART II

Evolving the Myths

CHAPTER 4

Robots Are Going to Replace (and Empower) Humans

A few years ago on a bicycle ride in Sag Harbor, I came upon a large truck on the side of the road filled with potatoes. I watched as a large machine raked through the field, plucked up the potatoes, and stacked them in the truck. As I thought of all the generations of people who took to the fields to dig through the dirt for dinner, I thought, *Wow, that's innovation at work!*

When we think of innovation, it's not a potato machine we typically think of, but that's a perfect example of how technology, that at first feels so bizarre, quickly becomes a part of our daily lives. Some people are afraid of technology, but we're not complaining that we don't have to go out in the backyard and dig up potatoes for dinner! The reality is that most things are automated now. And that's a *good* thing.

Technologies will continue to evolve, so it's up to us to face our fears and shift our perspectives.

A common fear about AI technology is that it will replace humanity. The reality, however, is that the robots aren't coming. In fact, they've been here for decades. Need proof? Look to your vacuum cleaner, toaster oven, car, garage door opener, and cell phone. Do you ever lose sleep worrying that your vacuum will take over your home while you sleep? Or that your toaster might come alive and usurp your dominance? Not likely.

Tech tends to creep into our lives, as it has over the last several decades, and suddenly we find ourselves unable to live without a blender, coffee maker, or smartphone. In a society that tends to fear change and a culture that seeks to induce fear, I want to do the opposite. My mission is to give people a sense of awareness of how technology unfolds—and how, if we give it a chance, it can join forces with us to make us a more evolved species.

The Robots Are Coming?

What is a robot? In simplistic terms, it is a machine that does something for you automatically. Even though such technologies are an accepted part of our daily lives now, it wasn't always that way. Like with many new technologies, there is an initial adoption phase when users' fears of the innovation interfere with the embracing of its benefits. It doesn't help this stigma that many of technology's most powerful players have continued to stoke such fears. Elon Musk, for example, the founder of Tesla and SpaceX, called AI humanity's "biggest existential threat" and compared it to "summoning the demon."[25] Yikes! British theoretical physicist Stephen Hawking told the BBC: "The development of full artificial intelligence could spell the end of the human race."[26] Gulp. No wonder we're scared! Though such fears are common in the general population, it is confusing to hear from technology's most powerful players, especially considering that Hawking—who suffered from a motor neuron disease—used technology to help him speak and move. Though these innovators are positioned to reap the benefits of the technologies they pursue, they have not accepted the challenge of educating people and thereby shifting perspectives.

Where do these suspicions of technology—and specifically robots—come from? And how valid are they? Like all worries, this one is rooted in the universal fear of the unknown. When we can't envision the future's innovations, it's hard to picture what our lives might be like in five, ten, or twenty years. This creates anxiety; for if we can't envision the future, how can we plan for it? This is certainly a valid worry and one that we won't easily dispel since it's largely innate and archetypal. The best chance we have to mitigate this fear of the unknown is by becoming educated. The more we know, the less we fear.

The Four Revolutions

What if you owned a farm that your ancestors had worked on for two hundred years, and one day a man in a suit turns up and says, "Hey, great news. I've got a tractor that can do the work of the entire family"?

---

25   Samuel Gibbs, "Elon Musk: Artificial Intelligence Is Our Biggest Existential Threat," *The Guardian*, October 2014, https://www.theguardian.com/technology/2014/oct/27/elon-musk-artificial-intelligence-ai-biggest-existential-threat.

26   Rory Cellan-Jones, "Stephen Hawking Warns Artificial Intelligence Could End Mankind," BBC News, December 2014, https://www.bbc.com/news/technology-30290540.

Your response might be, "Well, what are we supposed to do now?" In addition to being a major convenience, this would also be a major disruption. The son who was the great counter of sheep and cows on the farm could now be a data analyst. The daughter who tended and harvested the fields could now be a software developer. These advancements, however, are often overshadowed by the disruption itself.

With each of the four revolutions of innovation—the automation of farm workers, the automation of production workers, the automation of transaction workers, and the automation of knowledge workers—there were disruptions, and thus there were fears that accompanied them.

THE FOUR REVOLUTIONS:

1. Automation of farm workers
2. Automation of production workers
3. Automation of transaction workers
4. Automation of knowledge workers

The potential risks of labor disruption from automation have always generated anxiety. At the dawn of the Industrial Revolution—with its mechanized automations in manufacturing and agriculture and its steam engine—humans' fear of machines was born. The suspicions were grounded in logic, because machines were quickly changing the landscape of our workforce in profound ways. It seemed each innovation brought with it a nostalgia for simpler times; as if each time we took a step toward the unknown, we craved the safety of the known.

In 1900, 85 percent of the population was employed in primary, secondary, or tertiary agriculture.[27] The automation of farm workers completely reversed that stat, and now 15 percent of the population is employed in agriculture.[28] Thanks to the automation of farm workers, in the United States, the share of farm employment fell from 40 percent in 1900 to 2 percent in 2000, but Grandmother didn't need to work in the fields anymore to survive.[29]

Later, massive factories were built, with cities developing around them. When the second wave brought the automation of production workers, whole cities seemingly crumbled. In the United States, the share of manufacturing employment fell from roughly 25 percent in 1950 to less than 10 percent in 2010, but Grandmother didn't need to work on production lines anymore to survive.[30]

People then went and worked in supermarkets, airports, toll booths, and offices. When the third wave—the automation of transactional workers—hit the United States, 50 percent of all transactional jobs disappeared between 1972 and today, but Grandmother didn't need to work on the switchboard, or type letters for the boss, or hand out cash at the bank anymore to survive.[31]

At this point, people began working in accounting, managing, reporting, customer service, and advising. Now that the fourth wave is underway—the automation of knowledge workers—it is predicted

---

27 James Manyika et al., "Disruptive Technologies: Advances That Will Transform Life, Business, and the Global Economy," McKinsey Global Institute, May 2013, https://www.mckinsey.com/capabilities/mckinsey-digital/our-insights/disruptive-technologies.

28 Manyika et al., "Disruptive Technologies."

29 Manyika et al., "Disruptive Technologies."

30 Manyika et al., "Disruptive Technologies."

31 Manyika et al., "Disruptive Technologies."

to eliminate 50–60 percent of the work currently done by humans, but Mother (because Grandmother stopped working decades ago) doesn't have to be a clerk or a customer service agent.[32] Furthermore, doctors and people working in other specialized fields don't need to spend the majority of their time filling out reports, and experts don't need to spend nearly half of their time explaining to nonexperts what the data means.

Wave 1: The Automation of Farm Workers[33]

- In 1900, 85 percent of the population was employed in primary, secondary, or tertiary agriculture.
- The automation of farm workers completely reversed that stat, and now 15 percent of the population is employed in agriculture.
- In the United States, the share of farm employment fell from 40 percent in 1900 to 2 percent in 2000.

Wave 2: The Automation of Production Workers[34]

- The people worked in factories.
- Massive factories were built, and cities grew around them.
- The automation of production workers began to occur, and whole cities crumbled.
- In the United States, the share of manufacturing employment fell from roughly 25 percent in 1950 to less than 10 percent in 2010.

---

32 Manyika et al., "Disruptive Technologies."

33 Manyika et al., "Disruptive Technologies."

34 Manyika et al., "Disruptive Technologies."

Wave 3: The Automation of Transactional Workers[35]

- The people worked in supermarkets, airports, toll booths, offices, etc.
- In the United States, automation of transactional workers wiped out 50 percent of all transactional jobs between 1972 and today.

Wave 4: The Automation of Knowledge Workers[36]

- The people worked in accounting, managing, reporting, customer service, and advising.
- The automation of knowledge workers is underway and is predicted to eliminate 50–60 percent of the work currently done by humans.
- Doctors don't need to spend 52 percent of their time filling out reports.[37]
- And the experts don't need to spend nearly half of their time explaining to nonexperts what the data means.

Historical precedent shows the potential risk of disruption from automation has always generated anxiety. In all circumstances, however, while some jobs disappeared, new ones were created, although what those new jobs would be could not be ascertained at the time. Here is the statistical reality: For the past 120 years (apart from war time and the Great Depression), the average unemployment rate in first

---

35 Manyika et al., "Disruptive Technologies."

36 Manyika et al., "Disruptive Technologies."

37 Manyika et al., "Disruptive Technologies."

world countries has hovered around 6 percent.[38] The automation of knowledge workers is generating the same anxieties of the past, and yet it is clear that we stand at the dawn of a new renaissance. We are in the right place at the right time. We have the privilege of bringing to the world the most advanced technology the world has ever seen.

What's clear from history is that automations, technologies, and advancements are—and always have been—transforming everything for the better. All the previous innovation waves have been freeing human bodies from robotic physical labor. Arria's innovation wave, however, will dwarf all previous waves. Why? We are freeing human minds from robotic mental labor. The greatest renaissance in the history of humanity is already moving at full speed. The current confluence of enabling technologies, infrastructure, and market readiness indicates that there is a pending explosion of NLT adoption.

When we better understand how deeply rooted our fears are around the unknown, we better understand our own hesitation to embrace technological innovations. Of course we're wary! We've wrestled with humans versus machines for decades. Some fears are founded, and as scientists of the present continue to remind us, there is a balance to be struck between humans and machines. Our search for innovation can be both the thing that propels us forward and the thing that compels us backward. Like all things in the universe, we need to find the balance—between what is useful, miraculous, and what is arrogant, dangerous.

Once we comprehend how such innovations can free humans to do more and be more, we can better embrace the advancements of tomorrow. The irony is that many technologies—like natural language

---

38   International Labour Organization, ILOSTAT database, data retrieved on February 8, 2022.

automation—are created to bridge individuals with knowledge. Is it possible that in finding a balance with machine, we just might find our own humanity?

The Robots Are Already Here!

Just as the Industrial Revolution influenced people's daily lives in both wondrous and terrifying ways, the robotics revolution of the 1960s and 1970s has rekindled our fear of being overtaken by machines—more specifically, our *jobs* being overtaken. This fear has remained prominent as manufacturers have required increasingly sophisticated production chains and thereby required more sophisticated, powerful machines to build them. The automotive and manufacturing industries were early adopters of robotic technology, and the trend has grown rapidly in recent years with the manufacturing industry accounting for more than 86 percent of the world's stock of operational robots in 2016 and the automotive industry accounting for 43 percent.[39] In 2016, the share of robots in high-tech manufacturing grew from 21 percent in 2000 to 31 percent.[40] And in 2020, a record 2.7 million industrial robots—an increase of 12 percent—operated around the world.[41] Most of this stock has shifted toward new manufacturers in China, Japan, and the United States.

If human versus machine is an innate fear, then why has the robotic industry continued to grow in recent years? Why haven't we revolted at the usurping of our humanity by machines? Because outside

---

39 "How Robots Change the World," Oxford Economics, June 2019.

40 "How Robots Change the World," Oxford Economics.

41 "World Robotics Report 2020," International Federation of Robotics, press release, September 24, 2020, https://ifr.org/ifr-press-releases/news/record-2.7-million-robots-work-in-factories-around-the-globe.

of their benefits to manufacturers' productivity, there are proven benefits to humanity.

Recent data suggests that our innate fear of being replaced by robots is in fact misplaced because their value across the economy offsets their disruptive impact on employment. Though it is true that certain sets of workers have lost jobs to robots—and will continue to lose jobs—many other people in the wider population benefit from what Oxford Economics calls "robotics dividends"—stronger tax revenues, higher real incomes, and lower prices for manufactured goods. They found that in the manufacturing sector alone, a 1 percent increase in the number of robots per worker leads to a 0.1 percent boost to output per worker across the wider workforce.[42] They assert "that by displacing automatable jobs in manufacturing, robots free up many workers to contribute productively elsewhere in the economy, as they meet the demands generated by lower prices for manufactured goods."[43] Furthermore, their results show that a faster adoption of robots has positive impacts on both short- and medium-term growth. In fact, boosting robot installations to 30 percent above the forecasted baseline by 2030 would lead to an estimated 5.3 percent boost in global gross domestic product that year.[44] This is like adding an extra $4.9 trillion per year to the global economy by 2030![45]

---

42   "How Robots Change the World," Oxford Economics, June 2019.

43   "How Robots Change the World," Oxford Economics.

44   "How Robots Change the World," Oxford Economics.

45   "How Robots Change the World," Oxford Economics.

DIGGING DEEPER INTO ROBOTS

Combining the expertise of its specialists, Oxford Economics used the latest data to draw insights on the impact of robots worldwide in their 2019 report, "How Robots Change the World," which identified three main drivers behind the rapidly growing pace of adoption:

Trend 1: Price

Robots are becoming cheaper than humans. The report cites exponential growth in processing the power of microchips, extended battery lives, and larger networks as having lowered the average unit price of a robot.

Trend 2: Innovative Applications

As robot technologies advance, so do the capabilities of the robots themselves. Furthermore, they can be installed more rapidly and be used in more varied settings for more sophisticated processes. Since today's robots are smaller, more collaborative, and more in tune with their surroundings, robots have been adopted outside the traditional sectors of automotive and manufacturing industries.

Trend 3: Consumer Demand

Much of the demand for robots in recent years is due to the rising need for manufactured goods. As these requirements grow, some countries like China are rapidly investing in robots to position themselves as worldwide manufacturing leaders.

A robot can't take over your job, but they can assume your tasks. Technology causes a change in the *nature* of jobs, not in the *number* of jobs. In short, what could be automated should be automated.

It is naive to think robotics won't reshape the labor market just as mechanization did during the Industrial Revolution. What we must keep in mind, however, is that there are some proven benefits to this evolution of the market. What we have learned from the robotics revolution of the 1960s and 1970s is that automation and the use of industrial robots do replace some jobs, but they also create many others. When manufacturers increase their productivity levels, there is an increase in wealth, which likely results in job creation that offsets employment displacement.

Agentic AI technology will also decrease the need for those whose work is repetitive, where mundane tasks could be replaced by machines and increase the need for medium-skilled workers. With this shift, workers can expect to hold higher-paying jobs that highlight their uniquely human skills. Over time, the employers will think more about the inimitable human skills we offer to a job—like persuasiveness, interpersonal skills, creativity, and communication—rather than automatable skills. Understanding how to leverage those skills, and being able to work with technology rather than compete against it, will create new job opportunities in the workforce. Hopefully this shift will encourage a growth mindset that allows us to develop our skills rather than stay fixed and static. What a wonderful opportunity for human growth, and it's made possible by robots.

Believing that robots will impact humans in the workforce is only half the story because humans will equally impact technology. Humans empower AI by capturing subject matter expertise into software and giving it the power of language; AI empowers humans with machine-processing power to communicate advanced knowledge and understanding in natural language at machine speed—going beyond what humans alone are capable of. Together, the two automate mundane tasks to enable superhuman capabilities like advanced intelligence, peak performance, competitive edge, and total confidence. The full story is that the two empower each other: Hu + AI = superhuman.

Who doesn't want to imbue advanced intelligence, operate at peak performance, maintain a competitive edge, and operate with total confidence?

Case in point, human-written narratives, especially those based on data, are prone to errors and data quality issues. While some mistakes may go unnoticed or require little time to correct, others can be disastrous for an organization. This is particularly true if the mistake distorts key information for decision makers or clients, leading to huge amounts of rework and mitigation. Narratives generated by Arria, in contrast, don't face any risk of human error. In fact, a Forrester Research report that quantified the value of natural language automation found an 80 percent reduction in manual report generation labor and ROI of 209 percent.[46]

FINDINGS FROM FORRESTER: THE TOTAL ECONOMIC IMPACT OF ARRIA STUDY

- **Real-time insights:** A senior manager at a professional services organization said, "Feedback from leadership was that [Arria] is really valuable because they don't need to ask somebody all the time [for reports], and they get all the important information really conveniently presented there already."

---

46 "The Total Economic Impact of Arria Natural Language Generation Study," Forrester Research, 2021.

- **Time savings:** A senior manager in professional services explained, "Along the way, we would waste a lot of hours because one cell can drive a lot of mistakes. So [inaccuracy] was an issue, and it's definitely decreased since using Arria."
- **Decreased errors and rework:** An executive director in financial services described the benefits: "[Before Arria] it was turnaround time and overall time spent, and [then] somebody needed to really double-check it. Now we don't have to do that anymore."

Tech will create *more* opportunities for humanity, not take them away. Though technology may seem intimidating at first—as it did for the first generation of potato farmers who used machines in the fields or the first consumer to place bread in their newfangled toaster device—when we can shift our perspective to see the benefits to humanity, we are put at ease.

A study by the Economist Intelligence Unit that asked 502 senior executives from eight countries—Canada, France, Germany, India, Japan, Singapore, the UK, and the US—concluded that fears about machines encroaching on the workforce are greatly exaggerated. In addition to the operational gains cited by more than half of respondents, leaders also noted that automation created increased productivity, eliminated repetitive manual tasks, delivered tangible efficiency gain, increased revenue, and enhanced competitive advantages. Furthermore, a third of leaders claimed error reduction to be a significant benefit, and another 28 percent noted more reliable production processes.[47] Other benefits were also cited, including better customer engagement and more creative ways to find new sources of revenue.

Despite these obvious advantages, leaders still acknowledged the fear that many of their employees held about automation at the workplace. Scott Likens, the new services and emerging technologies leader at PricewaterhouseCoopers, offered, "I'm not sure the fear [of technology] ever goes away… But there's much more acceptance of the value that automation can add to the business. Managers are now saying, 'This is great—we can upskill our entire workforce.'"[48]

As these benefits trickle down to employees, we can expect to see more people adopting a growth mindset that allows them to enhance those skills that are uniquely human. Byron Reese, CEO and publisher of technology news claims that the human workforce will not be left behind if they are willing to ask the right questions: "What drudgery do I engage in that I can use technology to destroy? What new opportunities can technology give me that I didn't have before? Where can I buy back my own time?"[49]

Furthermore, considering automation's effect on the workplace, the study found that the skills that will be needed most by future workers include complex problem-solving, creativity, openness to change, collaboration, people management, intuitive thinking, more advanced education, emotional intelligence, and strong ethical values. As we can see from this list, these are uniquely human skills that cannot be replicated by AI. For this reason, we can rest assured that we cannot

---

47   The Economist Intelligence Unit, "The Advance of Automation: Business Hopes, Fears and Realities," UiPath, 2019, https://automationfirst.economist.com/wp-content/uploads/2019/06/EIU-UiPath-The-advance-of-automation-briefing-paper.pdf.

48   The Economist Intelligence Unit, "The Advance of Automation."

49   The Economist Intelligence Unit, "The Advance of Automation."

and will not be replaced by robots because they are incapable of the essential skills listed above. As long as we continue to develop and enhance those traits in us that make us more human, we will continue to thrive in the evolving workforce.

FIGURE 8: Given automation, what are the characteristics and skills believed to be most needed for the future workforce?
(% of respondents)

COMPLEX PROBLEM-SOLVING SKILLS
39

CREATIVITY
36

OPENNESS TO CHANGE
35

COLLABORATION
34

PEOPLE MANAGEMENT
27

INTUITIVE THINKING
22

MORE ADVANCED EDUCATION
22

EMOTIONAL INTELLIGENCE
20

STRONG ETHICAL VALUES
17

Figure 4.1. Source: The Economist Intelligence Unit

Technology will empower human actions just as much as humans will continue to empower technology. The two will continue to evolve simultaneously. Creativity is something that will always be uniquely human. Yes, we can use technology to augment what we do, but the ability to visualize, imagine, and dream up new things remains exclusively human.

Many people think the robots are coming. At Arria, we believe the robots are leaving. Too often in our hurried, divergent lives, we don't have time to invest in the things that make us uniquely human—like creativity, engagement, and meaningful relationships. It is because of technology that we are able to step away from repetitive, robotic tasks at home and at the office. At work, rather than pore over a large spreadsheet, an advisor has time to nurture clients, build networks and relationships, and be more creative. At home, rather than spend hours at the kitchen table doing complex math to balance budgets, a parent has time to spend with family, be creative, and enrich relationships.

Despite our innate fears about humans versus machines, the reality is that we can find a balance where technology can serve us in ways that make us more human. When we use technology to automate the robotic nature of the things that we do every day, we're freed to do more and be more. Our real fear is perhaps less about the robots and more about our own insecurities that we can find and maintain the balance between humanity and technology. Don't forget that you engage with technology to the extent that you choose. Just as Pandora's box didn't fly open of its own accord, neither will machines. We must hold the vision of a balanced future where machines continue to aid our humanity.

The most exciting potential of technology is in the ways it enhances humanity. Technology can truly change the world and allow us a more engaged and conscious role in it. We needn't fear the arrival of the robots because they are already here. Rather than fear their ability to replace us, let's celebrate their potential to empower us—to make us *superhuman*. In her poem, "The Summer Day," Mary Oliver asks the profound question: "Tell me, what is it you plan to do / With your one wild and precious life?" Thanks to technology, those plans can be limitless. The robots are not coming; they are leaving. Leaving humans free to create, discover, invent, build, heal, expand, relax, live, and love.

NOTES FROM AN UNWAVERING BELIEVER

Dr. Ehud Reiter, Founding Scientist of Arria

Twenty years ago, the UK Computing Research Committee asked me what my vision and "grand challenge" was for AI and NLG. I told them that I wanted NLG and AI to be used to humanize and explain data, so I gave the specific example of a tool that explained personal medical data. For example, if Mary Smith wanted to understand her health better, I would want to empower Mary with a tool that could analyze her medical records, extract key insights based on her concerns, and present those insights to her as an easy-to-read narrative, using language Mary could easily relate to.

I am convinced such tools are even more essential today. For example, there is now a massive amount of medical data available to the public, including information from sensors on phones and watches, but this data isn't useful if no one understands it. Doctors have also told me about cases where patients get stressed when viewing their health data because they spot something unusual and think there may be a problem, when all they are seeing is simply a natural variation and fluctuation in their medical history.

It is time to design tools that can summarize, explain, and contextualize data in a way that makes sense to Mary and the rest of us. This should also apply to all sorts of data, not just medical.

I would love to see natural language technology used to develop such solutions. Arria has developed revolutionary technology that extracts key insights from data and presents it as narratives to users. But the industry needs to build on this work to achieve this vision. We are required to work with massive amounts of data (such as medical records), which may include free text comments from doctors (etc.) as well as numbers and create narratives that work for people with varying levels of industry knowledge and literacy. We also need to make natural language systems interactive, so that Mary and other users can ask the right questions and focus on what they care about. Language systems should learn from their users, so they can get better at identifying and communicating key insights.

These are certainly some challenges, but I am also seeing real progress being made—both from universities (including my own research group at Aberdeen) as well as from companies producing products such as Arria Answers. I am convinced that we are on the right track to developing the universal system I dreamed about twenty years ago! It is exciting for me to watch this vision begin to take shape, and I am honored to be able to contribute to it, in a company that is helping to define the humanizing of data that works for everyone.

CHAPTER 5

AI Will Outsmart Us (and Make the World a Smarter Place)

Early in my career, I joined a start-up company called PaperDirect, founded by Warren Struhl. Warren was a great mentor and one of the most dynamic entrepreneurs I had ever worked with. When I started with the company, there were about thirty employees. Within three and a half years, there were five hundred employees, and the number of catalogs mailed and products created had grown exponentially. I first led the creative team that produced catalogs and designs for the laser printer papers and accessories sold around the world. I later became vice president and helped grow the business from a start-up to over $100 million in sales in less than four years.

It was thrilling to be a part of taking the company to the next level. When Warren brought in an outside agency to review the business, that, of course, meant a full review of me and my team. While I was highly recognized for my part in the company's growth, I was intimidated at first but quickly shifted my thinking. I realized this was an opportunity to learn and grow. With this mindset, no matter the outcome, I would have more knowledge.

After working together on the PaperDirect project, the outside team asked me to become a partner in their boutique Madison Avenue agency. This was a great lesson in getting over my fears quickly. In fact, it led to a partnership that went on to form Diligent and then Arria. This partnership ultimately took me around the world and taught me how to see numbers as a language and a spreadsheet as a creative road map.

The first time I was asked to design using a spreadsheet, my creative side cowered as I scrolled past rows and rows of cells. *How do I get creative with a spreadsheet?* My (now) business partner saw my apprehension and explained, "A spreadsheet is just a capture of data, and data is a language." In that moment, I realized there was indeed meaning in those cells; the numbers contained information. If I was unable to interpret them, however—if I didn't know how to speak the numbers' language—I couldn't access their value. If the data had a story to tell me, how could I hear it?

This moment stuck with me through my unlikely journey from my past in a creative field to my present in a technological one. I am a female in the tech world surrounded by experts and scientists who write algorithms that are changing the world. I absolutely love the sport of keeping up with them. I love analytics; I love creative. I like the balance that comes from using both sides of my brain each day. Most of all, however, I love helping data share its story.

It is no surprise that we have more information at our disposal than at any other time in history. But what does our unique position in history—with data swirling all around us—have to say about our humanity? What can we extrapolate from this data about what makes us uniquely human? And what story does the data share about our civilization?

The Datasphere

Our universe is filled with data. Neither the term "internet" nor the term "internet of things" is adequate to describe the magnitude, complexity, or importance of the vast global network to which the developer community has inextricably connected us all over the past twenty-five years. At Arria, we believe the term "datasphere" better recognizes the ever-evolving, dynamic, multidimensional, integrated nature of the "internet of everything": information, commerce, applications, social connections, devices, things, and experiences.

The datasphere is a multitrillion-dollar network infrastructure connecting billions of people via billions of connected devices and things. There are already more devices connected to the internet than there are people on the planet. Trillions of transactions flow daily through millions of interconnected applications and billions of interconnected websites and web pages. There are 2.5 quintillion bytes of data created each day at our current pace. But that pace is only accelerating with the growth of the internet of things. Over the last two years alone, 90 percent of the data in the world was generated.

Computers and machines continue to produce and store digital data at a rate that is impossible for human beings to make full use of. Despite these challenges, the data is economically perilous to ignore, especially if the competition is *not* ignoring it—and they're not! The volume of machine data is growing exponentially, and we are drowning in it. The problem, however, is not the data; the problem is that a datasphere—

which essentially contains the sum of all human knowledge flowing through it—cannot explain itself. Visualization is useful but not as effective as language itself. With forty-four zettabytes of data in the datasphere, how do you extract all the value that the data is presenting? Without an expert to analyze and explain it, the data is useless.

DIGGING DEEPER INTO DATA

How much data do we generate in a day? Considering that in 2025, there are 5.56 billion internet users, a number projected to increase exponentially in coming years, there is an unfathomable amount of digital activity happening each day.[50] In this vast digital ecosystem, a day in the data can mean the following:

- 500 million tweets are sent.[51]
- 376.4 billion emails are sent.[52]
- Five petabytes of data are created on Facebook.[53]
- Five terabytes of data are created hourly from each connected car.[54]
- 100 billion messages are sent on WhatsApp.[55]

50   Ani Petrosyan, "Number of Internet and Social Media Users Worldwide as of February 2025," Statista.com, April 1, 2025, https://www.statista.com/statistics/617136/digital-population-worldwide/.

51   Aditya Rayaprolu, "How Much Data Is Created Every Day in 2025?," TechJury.net, March 12, 2024, https://techjury.net/industry-analysis/how-much-data-is-created-every-day/.

52   Rayaprolu, "How Much Data Is Created Every Day in 2025?"

53   Matt Loy, "How Much Data Is Generated per Day?," DigitalSilk.com, updated December 13, 2024, https://www.digitalsilk.com/digital-trends/how-much-data-is-generated-per-day/.

54   Rich Miller, "Rolling Zettabytes: Quantifying the Data Impact of Connected Cars," DataCenterFrontier.com, January 21, 2020, https://www.datacenterfrontier.com/connected-cars/article/11429212/rolling-zettabytes-quantifying-the-data-impact-of-connected-cars.

55   Loy, "How Much Data Is Generated per Day?"

- 16.4 billion Google searches are made.[56]

The exponential growth of big data is difficult to perceive unless we can put it into perspective. So consider this: In 2020, there was a global volume of 44 zettabytes of data.[57] In 2025, that volume is 147 zettabytes—234 percent growth from 2020 just five years prior—and by 2028, it's projected that there will be 394 zettabytes.[58]

As data continues to grow, we need to brush up on our data literacy, so we are better prepared for the data-driven future.

ABBR.	UNIT	VALUE	SIZE (IN BYTES)
b	bit	0 or 1	1/8 of byte
B	bytes	8 bits	1 byte
KB	kilobyte	1,024 B	1,024 bytes
MB	megabyte	1,024 KB	1,048,576 bytes
GB	gigabyte	1,024 MB	1,073,741,824 bytes
TB	terabyte	1,024 GB	1,099,511,627,776 bytes
PB	petabyte	1,024 TB	1,125,899,906,842,624 bytes
EB	exabyte	1,024 PB	1,152,921,504,606,846,976 bytes
ZB	zettabyte	1,024 EB	1,180,591,620,717,411,303,424 bytes
YB	yottabyte	1,024 ZB	1,208,925,819,614,629,174,706,176 bytes

56   Anthony Cardillo, "How Many Google Searches Are There per Day? (March 2025)," ExplodingTopics.com, April 23, 2025, https://explodingtopics.com/blog/google-searches-per-day.

57   "A Day in Data," *Sunday Times*, n.a., https://www.raconteur.net/infographics/a-day-in-data/.

58   Khyati Hooda, "50 Data Generated per Day Stats to Know in 2025," March 17, 2025, https://keywordseverywhere.com/blog/data-generated-per-day-stats/.

Data is here to inform us, but we have to be able to understand it. Data, like language, is an incomplete process if there is no receiver of the information. If there is no way to interpret data, then why have it? As you might imagine, adding the power of language to every facet of the datasphere is a colossal undertaking. And that's where Arria steps in.

Capturing the Power of Language

Arria's patented language technology turns raw data into automated, expertly written reports in seconds, not hours or days. Arria starts working where current data analytics stop. It sits on top of big data and makes sense of it all. It makes data automatically communicate directly to you, not in numbers, spreadsheets, or visualizations that require further analysis and explanation but in rich narratives you would believe were written by a human expert. Because the Arria engine is an advanced AI software system, it can be replicated ten, one hundred, and one thousand times—doing the work of any number of experts 24-7, 365 days a year.

The Arria Language Platform enables an entirely new capability— the power of language—to be added to the datasphere. Prior to the advent of language automation, the automated reporting output of the analytics layer was limited to tabulated reports and visualizations. Despite these advancements, the automation of reporting was limited by the fact that both of these outputs require further expert analysis and explanation via human-authored written or verbal narratives. With the advent of natural language generative AI, the Arria platform expands the automated reporting output of the analytics layer to include not only tabulated reports and visualizations of the past but also natural language narratives of the future—without human intervention. These narratives describe in natural language not only the critical insights that were extracted from the data but also the meaning of the tabulated reports and visualizations. In fact, the narratives generated are so rich and compelling, you would believe they were written by a human who is a subject matter expert.

Likewise, adding the power of language to the datasphere is now rapidly gathering momentum, and anyone, anywhere can automatically deliver real-time, hyperpersonalized, expert, natural language narratives to users' devices—narratives that are indiscernible from human-authored text—with a natural language experience built and embedded into the tech stack to deliver the capability.

Arria systems completely change our relationship to big data, so we're empowered rather than overwhelmed by it.

It gives a voice to the data that, until now, has been trapped behind charts, graphs, and spreadsheets. It essentially allows the datasphere to talk to us—in a language we don't have to be experts to understand.

I remember when I showed Arria's comptroller demo summarizing data into long-form reporting or short-term presentations. She shared with me later that her first thought was, *Oh no, I'm going to lose my job!* After she experimented with it, however, she came to me and admitted that she hated writing reports in numbers. "I'm always afraid something will change at the last minute and change my report. Or that I miscalculated and the report is wrong based on human error." She realized that the technology wouldn't replace her job; it would merely change the nature of her job. Now, rather than sweating over long tabulations, trying to avoid mistakes, she has superhuman confidence and is freed to spend more time on other projects that require her uniquely human skills. Technologies such as Arria's bring the power of language to the most widely adopted business analytics tool, and its benefits are endless for finance, accounting, marketing, product management, and HR management sectors to use for day-to-day operations, business analytics, data compilation, and reporting. In fact, Arria can automate data analytics from many different industries, including quick-serve restaurants, consumer research, sports, real estate, government, pharma, energy, oil and gas, retail, consumer goods, and news and media.

Many people now assert that data is like the new oil because there is so much of it, and it is so valuable. From what I learned that day, poring over a spreadsheet I *couldn't* make sense of but *needed* to make sense of, there is no value in data unless we can extrapolate it. As Arria's founding scientist Ehud Reiter claims, "Arria gives a voice to the internet of things, bringing incomprehensible data to life." Business leaders understand data's value and know that it holds the information they need to access the health of their enterprise. All corporate leaders want to reduce costs and increase profits. How do you do that? Imagine having a company filled with expertise that drives the business forward but does not add costs. Unlimited access to data helps leaders recognize weaknesses in their business and thus how to rectify them. It also allows them to identify the allocation of their resources so as to redistribute them as needed to ensure they are used efficiently.

Despite the obvious benefits to the business sectors we've mentioned, Arria has broader implications for our culture at large. When data analysts capture the knowledge of a subject matter expert into software, they capture it without bias or judgment. This allows us a consistent delivery of information. Furthermore, because there is no human error, we don't spend time second-guessing ourselves because nothing is overlooked or miscalculated. We also don't get bogged down by data static that might prolong our processes and obscure the essence of the data.

Not all language technology is the same. Accuracy and consistency are key if we are to trust the data and the language that it generates. Trust is what will enable broad adoption and ultimately democratization. In this way, language platforms help rebuild our trust in data, information, and, in turn, ourselves. Furthermore, because the information relayed to us is based on data and facts, we don't need to question it or wonder if there is human error behind its calculations.

When capturing knowledge, we can harness the knowledge of top experts—not the best in our department or region but the best in the world. This raises the bar for all of us by bringing subject matter expertise to the masses. Suddenly everyone has the capability to view data the same way our best experts do. This is profound for corporate legacy planning. Suddenly you don't lose the knowledge of your brightest team members when they retire or transition. You can capture what they know and distribute it across departments or organizations to maintain consistency and ease changeovers. By embedding expertise into the engine itself, it becomes a purveyor of best practices in an organization that can be referenced again and again.

Though the human mind is our most powerful tool, we must admit its limitations. As Amir Husain, author of *The Sentient Machine*, explains, "Humans learn and forget. Computers can learn and don't have to forget. The notion of 'what is important' holds very different meanings in the context of a machine. We tend to remember more of what was important at the time, the machines can remember everything and then determine what part of this exactly preserved experience ends up being important at some later stage."[59] Using software to capture the breadth of expert knowledge—not merely what was important in a static place or time—we are expanding our own intellectual capabilities.

The larger implication for capturing expert knowledge unfettered by human limitations is that it makes the world a smarter place. When we all have access to experts' insights, then we are all more intelligent, and the world is smarter and more informed. With this advancement, we are freed to become more attuned, equal, engaged, and collaborative, and we free ourselves to evolve. Today, we can't imagine how we'd manage the web of text without search engines. In the very near future, we'll wonder how we ever managed big data without Arria. This technology continues to allow us to do more, be more, and know more to create an evolved world—like a Universe 2.0—made possible by using the power of language to bridge the gap between the growing datasphere and the human mind. Arria remains committed to bringing this technology to the world so we can bring forth the vision of a smarter, superhuman world.

---

59   Amir Husain, *The Sentient Machine: The Coming Age of Artificial Intelligence* (Simon & Schuster, 2017), 24.

CHAPTER 6

Technology Isn't Democratic (and Can Democratize Expertise)

If I'm an asset owner staring at a data set that contains fifty-four columns of information containing 1.2 million integers, it would take me hours to understand the analysis. In fact, there are entire teams and departments of asset managers that might spend days doing just that. Why are these columns of data important? Because the faster I can get to the information, the faster I can mitigate risks and seize opportunities. Now imagine a time when AI, modern technology design, and data all combine to give you, the singular asset owner, the knowledge of the top asset managers and experts in the financial community.

As we explored in the last chapter, with so much information available, it's hard to know where to start and who to trust. The first step in Arria's Language Platform is to take raw information and apply data analytics to extract meaningful facts: *What's significant? What's important in the data? What kinds of things could be talked about in that data?* The genius of Arria is that it assimilates, deciphers, understands, and provides written commentary in milliseconds on demand at the click of a button. In this first half of the process, we are essentially capturing information to present in a myriad of ways, like text or graphs. Next—and this is the secret science behind Arria's technology—we apply linguistics to the information to deliver it in articulate, fluent language.

Why is the knowledge *sharing* more significant than the knowledge *capturing*? Because in sharing information universally, expertise is democratized for all. No matter your demographics, industry, education level, or native language, you deserve the same information as the most specialized, affluent expert. Ultimately, Arria's knowledge sharing is designed for a more democratized world that promotes equality consciousness. This innovation marks a fundamental shift and illustrates how humans have the opportunity to evolve as they embrace AI.

Creating an Equal World Through Knowledge Sharing

With the datasphere growing wider and more varied each day, humans alone can't get information out fast enough before it's no longer relevant; superhumans, however, powered by AI technology, can. For many news organizations, this has created a novel dilemma: *How can news groups satisfy their audiences' demand for quality content in an environment with limited resources?* In partnership with Arria, BBC News Labs found a way to deliver rich, data-driven storytelling to their local news teams without substantially increasing their workload. As the world's leading public service broadcaster, BBC News uses Arria's technology to generate news stories at a superhuman, hyperpersonalized scale by taking local public data and turning it into news stories at a zip code level.

What BBC News and others have found is that by adopting automation into journalism, they are able to enhance their journalistic efforts. Robert McKenzie, editor of BBC News Labs, said, "This is about doing journalism that we cannot do with human beings at the moment … Using machine assistance, we generated a story for every single constituency that declared last night with the exception of the one that hasn't finished counting yet. That would never have been possible [using humans]."[60] Though some have feared that automated journalism will replace human journalists, BBC News has found that it has enhanced it. How? Because no amount of automation can replace the skill of structuring and relaying good stories. Furthermore, Arria's technologies enable writers to keep the same style that readers are accustomed to. McKenzie explains, "As a journalist, you try to think of every conceivable permutation of a story in advance, then you write a template. The machine selects particular phrases or particular words in response to precise pieces of data. So you can write everything if you want to, in 'house style.'" This marriage of humans and machines has enhanced the craft of storytelling because journalists no longer spend time analyzing the underlying data of their stories. Instead, they are empowered to offer readers greater ranges of stories that highlight what is uniquely human about the art of storytelling.

Even outside of the news industry, one set of information may be needed by various audiences. With natural language platforms, the same expert knowledge captured without bias can be shared in hyper-

---

60   Chris Fox, "General Election 2019: How Computers Wrote BBC Election
       Result Stories," BBC News, December 13, 2019, https://www.bbc.com/news/
       technology-50779761.

personalized formats for different recipients. Using the same shared data, reports can be drafted in different voices—one for management, for example, one for shareholders, one for board members, and one for customers. Where natural language automation is really good is where you have information from a lot of sources that needs to be combined, integrated, and presented to various audiences. There are a lot of places where people want the data wrapped up into a story. That's where generative AI is really powerful.

Being able to customize the way data is communicated for the needs of a particular audience is key to Arria's larger mission—to democratize expertise for equality consciousness. Just as no two people are the same, nor are their information needs. One of the starkest examples are the benefits within healthcare. During a health crisis, we are sometimes literally dependent on medical experts. It's a vulnerable feeling for the patient, and I'd imagine an overwhelming feeling for the expert as well. When a patient is in the hospital, data is being collected constantly about their condition, from vitals and labs to nurse notes and doctor reports. A key feature is the ability to analyze the delivery of that content to the specific audience receiving it. Now, with the touch of a button, multiple reports are automatically generated—one for the doctor with stats and high-level medical terms, one for the nurses to ease shift handovers, and one for the patients to educate them and provide actionable steps.

SAME INPUT. DIFFERENT REPORTS.

In an effort to investigate ways of presenting patient information to medical professionals and family members, the University of

Aberdeen created the BabyTalk research project. Their focus was on data in the Neonatal Intensive Care Unit.

Each of the reports below is driven from the same input data sources. NLT provides a customized, personalized, tailored information delivery, in the form of natural language narratives, automatically generated by machine, for different audiences from the same data set.

Doctor Report: Decision Support—NLT can produce reports for doctors, written in a language that doctors understand.

Current Status: At 1046 the baby is turned for re-intubation and re-intubation is complete by 1100 the baby being bagged with 60% oxygen between tubes. During the re-intubation there have been some significant bradycardias down to 60/min, but the sats have remained OK. The mean BP has varied between 23 and 56, but has now settled at 30. The central temperature has fallen to 36.1°C and the peripheral temperature to 33.7°C. The baby has needed up to 80% oxygen to keep the sats up.

Nurse Report: Shift Changeover—NLT can produce nurse shift handover reports, so that a nurse whose shift is ending can spend less of their time writing up a summary of what's been happening and more time attending to the child.

Current Status: Currently, the baby is on CMV in 35% 02. Vent RR is 50 breaths per minute. Pressures are 25/4 ems H20. Tidal volume is 5.7. Sa02 is variable within the acceptable range and there have been some desaturations down to 38. The most recent blood gas was taken at about 04:00. There is fully compensated respiratory acidosis

*or secondary compensation of metabolic acidosis. pH is 7.23. CO2 is
10.4 kPa. BE is 4.2 mmol/L. The last ET suction was done at 07:00.
There were large amounts of mucoid secretions. Oral suction was done.
There were large amounts of mucoid secretions. Currently, he is being
given 0.64 mls/hr of morphine via contin-uous infusion. He is on
CPAP in 35% O2.*

*Events During the Shift: The baby was intubated at around 06:45 and
was on CMV. Vent RR is 50 breaths per minute. Pressures are 25/4 ems
H20. Tidal volume is 5.7. FiO2 was lowered to 35%. Since around
07:45, he has been on 0.64 mls/hr of morphine via continuous infusion.*

Parent Report: Status Updates—NLT can produce, from the
same input data sources, an appropriately worded letter to the
parents of the child, reassuring the parents about the child's state
of health.

Baby John

Date of Birth: 05 Jan 2012

During the day, Nurse Johnson looked after your baby.

During the night, Nurse Smith looked after your baby.

*Your baby, John, is receiving intensive care at the Royal Infirmary of
Edinburgh. He is being looked after in Blackford nursery in cot space five.*

*John is now 2 days old with a corrected gestation of 24 weeks and 2
days.*

*His last recorded weight is 460 grams (1 lb 2 oz). Because John was
born earlier than expected, he has been nursed in an incubator. This*

keeps him warm by keeping the heat and humidity in the incubator and preventing him from losing too much moisture from his fine skin.

John is currently receiving ventilation support. Ventilation helps to provide the support that enables him to breath more easily. It does this by giving extra breaths, pressure, and/or oxygen to baby's lungs. So that your baby's lungs remain open for oxygenation. In the morning, the amount of oxygen required for your baby was around 27%. In the last 12-hours this has been between a high of 50% and as low as 27%.

Baby John has been administered the drugs Morphine (Analgesic) and Suxamethonium.

Source: Neonatal ICU, Royal Infirmary, Edinburgh, UK. Included with permission from the University of Aberdeen.

Clearly the narrative we need to provide each recipient should be different. NLT makes mass customization possible. The robotic dividends addressed in chapter 4 apply here as well: Without the use of human labor on reports, doctors are freed to spend more time with their patients; nurses can build relationships and offer support; and patients can spend less time feeling afraid and more time feeling informed and thereby more empowered for healing. Sounds superhuman, right? It is. Allowing technology to assume our more rote behaviors frees us to do more of the things that make us human. After all, the greatest power of the machine is that it makes humans more human.

The Wider Impact of Language Automation on Culture

Much of our culture's unease is a result of power imbalances. With the growing divide in wealth and education, there is an inequity of power that can be both helped and hindered by technology. Innovations like language automation are attempting to bridge that divide. **Danah Boyd**, principal researcher at Microsoft Research and founder of Data & Society, acknowledged this divide in response to Pew Research: "Democracy requires the public to come together and work through differences in order to self-govern. That is a hard task in the best of times, but when the public is anxious, fearful, confused or otherwise insecure, they are more likely to retreat from the collective and focus on self-interest. Technology is destabilizing. That can help trigger positive change, but it can also trigger tremendous anxiety."[61] Boyd recognizes that technology can be used to disempower us and empower us. It causes imbalances in power that can benefit a few or benefit the masses.

Technology is a mirror for humanity. With the advent of social media specifically, the good, the bad, and the ugly are amplified tenfold. When a minority of people own the majority of knowledge, it creates intellectual inequality that results in limited opportunities and diminished possibilities. With technology, we can open up the channel of understanding by democratizing expertise. When this is reflected in our technology, it is also seen in our culture's psyche. With an evolved equality consciousness, every person has an equal level of understanding. Information is now available to everyone, regardless

---

61   Janna Anderson and Lee Rainie, "Many Tech Experts Say Digital Disruption Will Hurt Democracy," Pew Research Center, February 21, 2020, https://www.pewresearch.org/internet/2020/02/21/many-tech-experts-say-digital-disruption-will-hurt-democracy/.

of age, gender, industry, education, language, or region. When our technology evolves, so do our people.

Humans need information in different ways, for different reasons. Wherever there is data—from news to financial reports to clinical safety reports—there is a need to deliver personalized content for users. Arria's innovations mark an evolution from generalization to personalization. It's true that Arria makes industries more efficient and accurate, but those are not our primary goals. Our mission continues to keep humanity at the heart of technology. So what are the benefits to humanity? As **Michael Wollowski**, associate professor of computer science and software engineering at Rose-Hulman Institute of Technology, succinctly claims, "If citizens cannot form an unbiased opinion, then democracy is lost."[62] NLT provides unbiased information to recipients, which not only helps them make more informed decisions, but on a global level it restores democracy and empowers humans.

We must not forget that technology is a microcosm for culture at large. **David J. Krieger**, director of the Institute for Communication & Leadership, based in Switzerland, explains the profound effects that democratizing expertise in the digital world can have in the physical world: "The digital transformation supports values such as communication, participation, transparency, the free flow of information, connectivity and authenticity. On the basis of these values, democracy will become more responsive to citizens, who will be able to access more information, assess the value of information and participate in shaping and using information."[63]

---

62 Anderson and Rainie, "Many Tech Experts Say Digital Disruption Will Hurt Democracy."

63 Anderson and Rainie, "Many Tech Experts Say Digital Disruption Will Hurt Democracy."

As we've discussed, many of the fears about technology and AI are effects of being uninformed. Fear stems from a lack of information, not from a lack of true understanding. How do you rid yourself of fearing the unknown? Make it known. Without understanding, we can often feel disempowered, afraid, uncertain. When we understand that AI can democratize our culture, we begin to live with less fear and more agency. If we let it, technology can humanize us, advance us as a species, and strengthen what is uniquely human about us. All of a sudden, we become more equal, more superhuman, and, in turn, the world becomes more evolved.

I am an unwavering believer in this vision of an evolved world made possible by democratizing expertise and understanding. **Paul Saffo**, chair of future studies and forecasting at Singularity University and visiting scholar at Stanford MediaX, once stated that new technologies are often met with initial suspicion before being integrated as positive forces in our culture. He says, "New technologies are like wild animals—it takes time for cultures to tame them."[64] Once they are "tamed," or fully understood and implemented, then the human race can benefit from them. Though I understand his point, new *language* technologies are different, as they are inherently more familiar and uniquely human. Witnessing the output of language technologies creates a different user experience since the medium is one we have communicated with for centuries.

Language is not just a communication tool, a problem solver, an information vehicle; it's also a way of providing comfort, building relationships, and expressing ourselves. We want the vast collection of information in the datasphere to empower us. With natural language

---

64 Anderson and Rainie, "Many Tech Experts Say Digital Disruption Will Hurt Democracy."

technologies, we are no longer waiting for experts to personalize troves of information, and the experts themselves are now doing bigger and better things. So what happens next? We evolve. We keep pushing the boundaries, leveling the field, and raising the bar. We become superhuman.

NOTES FROM AN UNWAVERING BELIEVER

Kevin Kelly, Author and Founding Executive Editor of Wired Magazine

Adapted from an official TED conference in February 2005

I don't know about you, but I haven't quite figured out exactly what technology means in my life. I've spent the past year thinking about what it really should be about. Should I be pro-technology? Should I embrace it full arms? Should I be wary? ... So I'm still perplexed about what the true meaning of technology is as it relates to humanity, as it relates to nature, as it relates to the spiritual.... Technology is accelerating evolution. It's accelerating the way in which we search for ideas.... And I believe that technology is actually a cosmic force.

The origin of technology was not in 1829 but was actually at the beginning of the Big Bang, and at that moment the entire huge billions of stars in the universe were compressed. The entire universe was compressed into a little quantum dot, and it was so tight in there, there was no room for any difference at all. That's the definition. There was no temperature. There was no difference whatsoever. And at the Big Bang, what it expanded was the potential for difference. So as it expands and as things expand, what we have is the potential for differences, diversity, options, choices, opportunities, possibilities and freedoms.... That's what technology is bringing us: choices, possibilities, freedoms. That's what it's about. It's this expansion of room to make differences.

And I think it's really important. Because if you can imagine Mozart before the technology of the piano was invented, what a loss to society there would be. Imagine Van Gogh being born before the technologies of cheap oil paints. Imagine Hitchcock before the technologies of film. Somewhere, today, there are millions of young children being born whose technology of self-expression has not yet been invented. We have a moral obligation to invent technology so that every person on the globe has the potential to realize their true difference. We want a trillion zillion species of one individual. That's what technology really wants.

Our humanity is actually defined by technology. All the things that we think that we really like about humanity are being driven by technology. This is the infinite game. That's what we're talking about. You see, technology is a way to evolve the evolution. It's a way to explore possibilities and opportunities and create more. And it's actually a way of playing the game, of playing all the games. That's what technology wants.... Every person here has an assignment. And your assignment is to spend your life discovering what your assignment is. That recursive nature is the infinite game. And if you play that well, you'll have other people involved, so even that game extends and continues even when you're gone. That is the infinite game. And technology is the medium in which we play that infinite game. And so I think that we should embrace technology because it is an essential part of our journey in finding out who we are.

CHAPTER 7

The Potential of Artificial Intelligence Is Terrifying (and Exhilarating)

As a technology innovator, I'm used to pushback. Though you might think people are always excited to use the next best and brightest innovation, it rarely happens that way. When my partners and I were building Diligent Corporation—an SaaS company that pioneered the board portal in the early 2000s, we knew there was a need to automate the process of compiling, managing, and archiving board meeting materials. After all, our software allowed users to replace the paper board books they had been lugging around for decades, gave directors more time to read materials, and reduced the high labor cost of a general counsel who prepared the materials. This all seemed so obvious. What we didn't realize was that we were disrupting the way board meetings were traditionally managed, and some clients were not as eager to accept this innovation. We quickly noted that for some clients, their hesitance was not because of the software itself but because of the changes it brought.

People are afraid of change. It's human nature to get comfortable with "what is" and fear "what might be." In 2007, Apple announced its iPhone, which it claimed ushered "in an era of software power and sophistication never before seen in a mobile device, which completely redefines what users can do on their mobile phones."[65] Knowing what we do about the prevalence of smartphones in today's world, we might assume people were scrambling to be one of the first users of a technology that would reinvent the mobile phone. But that's not exactly what happened.

Here's how NPR described this newfangled device called an iPhone: "It's a cell phone, it's a music player, it's a camera, it's a web-enabled device, and much more. Ask yourself if you really need all that high-tech bling."[66] And according to *The Guardian*, "Apple's much-anticipated iPhone, which goes on sale in the US today, will struggle to break into the mainstream because of a lack of a 3G connection and low demand for converged devices, according to research."[67] *New York Magazine* did acknowledge the product's innovation but still doubted its—and its creator's—survivability: "It's a stunning box, a wizard object with a passel of amazing features (It's a phone! An iPod! A web browser!). But for all its marvels, the iPhone inaugurates a dangerous new era for Jobs.

65   "Apple Reinvents the Phone with iPhone," Apple Newsroom, press release, January 9, 2007, https://www.apple.com/newsroom/2007/01/09Apple-Reinvents-the-Phone-with-iPhone/.

66   Maria Godoy, "Seven Things to Consider Before Buying an iPhone," NPR, June 27, 2007, https://www.npr.org/templates/story/story.php?storyId=11449714?storyId=11449714.

67   Jemima Kiss, "iPhone Set to Struggle," *The Guardian*, June 29, 2007, https://www.theguardian.com/media/2007/jun/29/digitalmedia.news.

Has he peaked?"[68] Of course we all know that Jobs did not, in fact, "peak" with the iPhone, and the Apple company has continued to push innovation—whether customers are welcoming of it or not.

The real tipping point for Apple came with the release of the iPad. Suddenly we had a fun, cool device that ultimately became a new standard. It was this moment that was an inflection point for the field of technology innovation as a whole. Consumers learned that innovation—and the changes it brings—can actually make our lives more fun, efficient, relaxed, and easy. Did the technology evolve, or did we evolve? Both. It's hard to distinguish which happened first, but there is no doubt that innovations like the iPhone propel our culture and, in turn, our future innovations.

As technology innovators, when we figure out how to make a product about the *experience* itself, we can mitigate the fear of change. In the same way, AI can be scary because change is scary, but when it becomes an experience that advances our way of life, it's no longer intimidating. Technology might be disruptive in the short term, but in the long term it changes us for the better.

Let's return to the mechanized factories powered by steam engines in the Industrial Revolution. The microview is that the modern production methods of the time did change the workplace and replaced some jobs. The macroview, however, is that the methods revolutionized manufacturing and led to other innovations to come. This doesn't happen overnight, however, as Siemens US CEO Barbara Humpton reminds us: "The context I like to keep in mind is when companies went from steam to electric, they tended to electrify the production line exactly as it had been set up in its steam configuration. Then the

---

68   John Heilemann, "Steve Jobs in a Box," *New York Magazine*, June 15, 2007, https://nymag.com/news/features/33524/.

groundbreakers said, 'Wait, I can put equipment in any configuration I want because I don't have to power it down a steam production line.' But the thing we're talking about a lot here at Siemens is, with each one of these changes, we've elevated the role of the human in the process."[69] Even at this moment, she adds, we are seeing such influences at work: "The fear that something is changing and the awareness that, as we do this, we unleash people to be doing things that add value in new and different ways."[70]

Considering that evolution is the process by which species adapt over time in response to their changing environment, we can start to see the larger opportunities—and our larger role—in today's technological innovations. Technology has evolved, and now we need to evolve alongside it. As Amir Husain, author of *The Sentient Machine*, writes: "With knowledge, we can self-direct our own conventional, biological evolution and collaborate with artificial intelligence in the most profound ways. The most famous horses are the ones that run the fastest; the most impressive elephants are the ones with the longest tusks; and the most notorious snakes are the ones with the deadliest venom. In a similar way, the greatest of our humanity is that which is able to self-evolve."[71]

As AI gets better at replicating the human mind, there is an important paradox at work here: Technology is essentially trying to keep up with the human brain, while the human being is trying to keep up with technology. As technology evolves intellectually, we are called

---

69   C. J. Prince, "Can Robots Replace Humans? Just Ask Elon Musk," Society for Human Resource Management, October 5, 2018, https://www.shrm.org/resourcesandtools/hr-topics/technology/pages/can-robots-replace-humans-just-ask-elon-musk.aspx.

70   Prince, "Can Robots Replace Humans?"

71   Amir Husain, *The Sentient Machine: The Coming Age of Artificial Intelligence* (Simon & Schuster, 2017), 181.

to evolve emotionally, or "self-evolve" as Husain calls it. As with all paradoxes, there is a balance to be struck. The balance here is that technology is advancing us forward—if we let it.

At no time has this paradox been starker than during the global pandemic of 2020. Almost overnight, we were rethinking how we might stay connected to our colleagues, teams, clients, friends, families, and communities. People needed to do more with less because of the unforeseen global disruption, and technology acted as that bridge. We had Zoom weddings, virtual Thanksgivings, FaceTime happy hours, and texting capabilities—all thanks to technology. Had we not evolved enough to invent technology, we would have had no way to remain connected with other humans, and our own individual mental health would have suffered, as would our collective evolution. As author Peter Thiel writes in his brilliant book *Zero to One*: "Humans are distinguished from other species by our ability to work miracles. We call these miracles *technology*."[72]

Innovation is both terrifying and exhilarating at the same time, but this is precisely how we advance our species forward. Once we get over the initial fear of disruption, we can see that technology is easing our evolution into the unknown future. Even though we couldn't have predicted the global pandemic, for example, technology was ready for it. We're afraid until we're reliant on a technology, and then our fear turns to gratitude for the many miracles it affords us.

As more and more disruptions happen in the modern world, we will continue to turn to AI to bridge divides. As Arria's founding scientist Ehud Reiter explains, "There are certain things people are really good

---

72   Peter Thiel, *Zero to One* (Crown Publishing, 2014), 2.

at and certain things computers are really good at."[73] Where machines can do routine work quickly while maintaining limitless records of data used, humans can synthesize at high levels. As Reiter points out, "Ideally, you want to combine the two."[74] It's a profound responsibility to ask technology to do what humans are uniquely capable of. Why do we do it? Because the rewards make us more human—more superhuman, like the ability to communicate, connect, build relationships, advise, and inform.

Rather than fear each new innovation for the change it brings, we need to understand the larger pattern of humans' evolution alongside technology. One technology leads to the next and the next: Tabulation led to visualization, led to automation, led to narration. A visual might be persuasive, but it is more so when supported by words, preferably spoken words. After all, charts and graphs can't express value propositions, establish thought leadership, or give a compliment. Furthermore, the automation of robotic processes and the written expression of those insights in natural human language are steps in the right direction, but to make the process feel even more human, we have evolved to language.

Our most important communications take place in natural conversation. Conversation has been the default mode of communication for millennia. What if, instead of fingers on the keyboard and eyes on the screen, we could simply have a natural conversation? Not a command-response "conversation" of the kind that we are accustomed to having with voice platforms such as Siri, but a true two-way conversation with platforms such as ChatGPT, during which the platform is not merely responsive to simple questions but is spontaneously analytical,

---

73   S. A. Mathieson, "Natural Language Generation Progresses from Robo-Journalism to Finance," *Computer Weekly*, November 27, 2015, https://www.computerweekly.com/feature/Natural-language-generation-progresses-from-robo-journalism-to-finance.

74   Mathieson, "Natural Language Generation Progresses."

accurate, and a subject matter expert in your business and field. What if this experience included the ability to bring to your attention critical insights discovered within the data so that you can make better, faster business decisions?

Arria's conversational-AI offering, called Arria Answers, makes it easier to analyze data. It gives users real-time access to key insights from data—using natural, spoken language. A true breakthrough! Arria Answers brings together Arria's foundational experience in NLG and insight analytics with other NLTs like NLP, NLU, and NLQ. This combination delivers true multiturn drill-down capability to maintain the conversational context, which makes talking to a digital assistant fluid and natural. As Arria's chief technology officer Neil Burnett puts it, "With Arria, you get the data insights spelled out for you in plain English by software that has no personal agenda. So the interpretation of critical information is objective, and everyone has the same take-aways. All your dossier viewers are equally empowered to identify, communicate, and action key insights."

The low level of effort required to have a conversation with voice platforms will accelerate the democratization of data literacy across whole organizations, leading to broader knowledge, greater collective intelligence, and ultimately better, faster business decisions. It will also improve the workplace environment by making data analysis and reporting more enjoyable. To some degree, it may also untether employees from their keyboards and screens. We have come full circle to a state where natural language—which has long been the preferred mode of communication among people—will be established as the ultimate communications interface between humans and machines. Is it any surprise that what's next is a call back to where it all began? That's the power of language.

A Shared Future of Intelligence

As we discussed in chapter 4, what could be automated should be automated. Sure, a human can sift through pages and pages of data, sorting the static from the insights, but should you? Once you are finished, will you doubt your findings? Worry you missed something? Without technology, humans spend a lot of time looking behind them at what they overlooked or miscalculated; with technology, humans are free to look ahead—to dream, invent, innovate, inspire. Perhaps the greatest gift technology offers us, and will continue to offer us, is the opportunity to expand ourselves beyond where we currently are. But what if we refuse to evolve? What if stasis is so comfortable that we cannot move even when given the opportunity?

For every unwavering believer like myself, there is a staunch unbeliever. Given my core philosophy that everything is a paradox, however, this doesn't discourage me. In William Shakespeare's *As You Like It*, the melancholy cynic Jacques muses, "All the world's a stage, / And all the men and women merely players." People play different roles, and each has a purpose. As with any paradox, though, some roles seem to contradict each other; on closer examination, you can't have one without the other. Believers and doubters need each other—not in some figurative sense but in a literal sense. One cannot believe in something unless it can also be doubted. In physics, this is called the law of polarity. According to universal law, everything has a polar opposite. Hot must have cold, up must have down, joy must have sadness, life must have death, day must have night—for without the other, we could not discern a difference between the two. Even forces are paired in opposites as the law of motion states—for every action there is an equal and opposite reaction.

There is dualism in all things, and the same is true when it comes to believers and unbelievers in the shared future of humans and machines. We often think of such dualities as being in conflict, but that is a limited perspective. When we can see with a broader lens, we understand there is a larger story to behold outside of our own limited role. Some ancient philosophies encouraged humans to maintain this larger perspective, one of the most famous being the Chinese concept of yin and yang. The image of yin and yang shows how seemingly opposing forces are actually complementary, interdependent, interrelated, and interconnected. When we can see outside our limited roles and grasp the mutual whole, we are growing and evolving. Within every innovation, within every setback, there is potential for both success and failure—because one is needed in order to validate the other.

Humans' role then, as innovations continue to emerge at a rapid pace, is less about identifying our individual role and more about understanding the collective whole. For example, I am an unwavering believer in innovation and its potential to propel human evolution, but I spend much of my time entertaining the opposite, for I cannot believe if there is no doubt. In order to instill equality consciousness, as is Arria's larger mission, we have to think this way. As the founder of analytical psychology, Carl Jung, reminded us, "There is no energy unless there is a tension of opposites; hence it is necessary to discover the opposite to the attitude of the conscious mind."[75]

Our culture needs the visionaries as much as it needs the skeptics, and part of equality consciousness is understanding the dual nature of things. When we can appreciate the beliefs of others, truly appreciate them as gifts that allow us to hold our own beliefs, then we are congruent with

---

75   Carl Jung, *Two Essays on Analytical Psychology* (Routledge, 1999), 53.

universal law. As seen in the natural world, when we are in our most evolved state, we are not in conflict—we are in concert.

DIGGING DEEPER INTO DUALISM

At the root of some ancient philosophies lies the concept of dualism—or the inherent contrasts that appear throughout the natural universe, like light and dark or push and pull. Such polarities are not to be confused with conflict. In other words, it's not light versus dark; it's light interconnected with dark. In these philosophies, humans can learn from the law of polarity that we see at work in the natural world around us. Even Albert Einstein's theory of relativity demonstrates this dualistic principle since time and space are not separate but depend on each other. The principle of polarity holds that differing aspects—like light and dark—are not two separate systems but are instead two parts of the same system. In fact, without one the system crumbles.

In Chinese philosophy specifically, cosmic energy is seen as having two principles: yin (negative) and yang (positive). The goal is not to banish one in service to the other but instead to keep the two in balance. After all, there cannot be one without the other. This balance is called *hsiang sheng*, meaning inseparability, or mutual arising as the ancient Chinese philosopher Lao Tzu wrote in *Tao Te Ching*:

When everyone knows beauty as beautiful,
there is already ugliness;
When everyone knows good as goodness,
there is already evil.

"To be" and "not to be" arise mutually;
Difficult and easy are mutually realized;
Long and short are mutually contrasted;
High and low are mutually posited;
Before and after are in mutual sequence.

If so many concepts in the natural world display these polarities, why do they seem separate to us? According to Chinese philosophy, it's because our human ego makes things appear separate. When we consider that astrophysicists now estimate that 90 percent of the material in the universe is invisible, we start to better understand these ancient philosophies and how they can provide us a mindset that is still relevant in our modern world. The call to action seen in these ancient traditions that is still relevant today is to rise above the ego to see that all seemingly contrasting aspects are, in fact, two essential parts of the whole. Armed with a mindset that sees the whole perspective rather than warring parts, humans find peace.

The Future: Humanity at the Heart of Technology

Once I integrated the universal concept of dualism into my thinking, I stopped wasting time trying to change people's minds about technology and instead spent more time trying to step into people's minds. Change is scary, and as such, AI can feel terrifying. The automation that changed our world forever sought to replicate human actions, but with AI we are working to replicate human thinking. If the story stops there, then it is scary! But the story keeps going because humans keep evolving. It's not just about what is lost with AI. It's also about what is gained—time, freedom, expertise, knowledge, equality.

As we saw with the creation of the internet, having information at our fingertips that we can understand in plain language empowers the world. That doesn't mean, however, as Apple learned with the launch of the iPhone, that it is immediately welcomed by the masses. Ultimately, despite our fervor for innovation or our fear of it, we must surrender to the paradox. We will never understand everything. We can have tenfold the data we have today, and that doesn't suddenly make us prescient beings.

In a letter to his brother in 1878, the famous painter Vincent Van Gogh wrote of the human struggle that uncertainty brings:[76]

> There was once a man who went to church and asked, 'Can it be that my ardor has deceived me, that I have taken a wrong turning and managed things badly? Oh, if only I could be rid of this doubt and know for certain I shall come out victorious and succeed in the end.' And then a voice answered him, 'And if you were certain, what would you do then? Act now, as if you were certain and you will not be disappointed.' Then the man went on his way, not unbelieving but believing, and returned to his work no longer doubting or wavering.

In order to live the full expression of our human lives, we must "act now" as if we're certain. The only way to do that is by living with agency. We must take responsibility for our own gaps in understanding; we must become "certain." This is the way forward for individual evolution. Every day, live the principle: *Seek first to understand.* The more information we have, the more aware and engaged we are, and more importantly the more peace we feel. If we all committed daily to becoming informed about which we are uninformed, the world

---

76   Vincent Van Gogh, personal letter, April 3, 1878, https://www.vangoghmuseum.nl/en/highlights/letters/143.

will be an exhilarating place, no longer "doubting or wavering." This is the power of knowing.

At Arria, we believe that the path to human understanding is paved by moving the power of language from the mind to the machine. In doing so, we are igniting a global revolution in communication. We envision the natural language tools and systems we are bringing to the world will fundamentally change the way people interact with technology, making us more informed to go forth and create a perfectly exhilarating paradox of believers and unbelievers.

Natural Language Technology: More Than a Chatbot

Chatbots, now known as "agents," have become a standard part of customer service and marketing for many businesses. Like all AI solutions, there are missteps in the beginning, but eventually the technology becomes smarter. At Arria, we work with natural language generation, a subset of generative AI and agentic AI that lends itself to smarter conversational interaction rather than rote, programmed answers to presumed questions.

The natural language technology that Arria uses in its chat-enabled platform is based on proactive conversational AI. It delivers answers that are context-based and anticipatory. When a system knows the context of the conversation you are having with your virtual digital assistant (VDA), it can correctly identify and anticipate your real intents. By contrast, if the system operates under a reactive model as chatbots do, it treats the digital assistant like a basic query system.

When you think of conversational AI, you likely think of chatbot agents. AI agents are often beloved by businesses for the quick customer service response they allow for and often detested by customers. They allow the business to be responsive at a lower cost than a human helpline, and sometimes they work well for customers. Often, however, they can only deliver surface-level information, too often not the information the customer is seeking.

But what about conversational AI that delivers an altogether different experience and is more than a first-line customer service solution? What about conversational AI that allows you to engage with your data more easily and more quickly get answers to the business questions you have? Say you want that.

And if you want that, do you also want it to employ proactive conversational AI? What does that even mean?

To understand the difference between the two and why natural language technology is superior to chatbot technology, let's look at what chatbot technology is.

Like NLT, chatbots are built on AI technology, allowing the tech to "chat" with you. A chatbot can answer your questions, to a degree, and provide information based on preset answers to questions the chatbot expects you to ask. There are chatbots designed to help you with customer support, but they are not insightful.

Think of the difference in language evolution as a child grows. The conversation you have with a four-year-old is usually to deliver information to the child. The questions they ask are typically easy to answer, and you have the answers at the ready.

Now, think of the conversation you may have with an adult, a hyper-intelligent person, possibly a data scientist or mathematician. The questions they ask may leave you searching for answers, or unable to answer because the conversation is anything but predictable. Consider how frustrating it is when a chatbot delivers a "predictable" answer to a question you did not ask.

What Arria brings to the conversation is another level of human interaction and adeptness; AI erases much of the frustration chatbots have become known for. Arria generates answers to your questions by connecting ideas and layering in additional information. Arria's natural language makes you think harder, think about something else, something beyond your question.

When it comes to Arria, conversational AI that is proactive means a system that does not just answer your question, but it answers the next question you are going to ask too.

What does the move from chatbots to proactive conversational AI look like in business? Say you ask your AI agent, "How are my sales?" The AI agent might tell you, "Sales are $1.57 million." There. You have been given a simple fact. No context, no causes.

When the answer goes further, to anticipate follow-up questions and answer those too, then you are getting proactive conversational AI: "Sales are $1.57 million, in line with your target" or "Sales are $1.2 million, a decrease of $1.3 million from the same period last year." When conversational AI goes further, it helps guide you through your data. Conversational AI that is proactive can identify the insights and highlight potential problems or wins for you, so you don't have to go digging around looking.

Arria's focus in the sphere of conversational AI is building proactive systems. Our conversational-AI offering, called Arria Answers, is a platform that highlights key insights without the user having to repeatedly drill down. It can be integrated with any existing platform, giving companies real-time access to business insights—delivered using natural, spoken language.

When the AI agent proactively solicits input, it can provide insights more quickly, it reduces ambiguity, and it can lead to better-informed decision-making. Proactive conversational AI gives you confidence that performance indicators are on track without having to query your data over and over.

With Arria Answers, we are bringing to AI agents our foundational experience in natural language generation—along with our innovative insight analytics. Arria Answers offers machines the power of natural human language, empowering a digital assistant to be smart, contextual, and accessible.

Smart. Arria Answers anticipates what you want to know. It enables your AI agent to answer your next question before you even ask it. An AI agent needs domain knowledge to deliver key insights and metrics without reporting every little fact that might be uninteresting or unimportant. It needs to remember and understand your preferences as an individual rather than responding in the same way to each user. And it needs to understand synonyms, idioms, slang, acronyms, and other complex language features that allow you to interact in a truly human way.

Contextual. Arria Answers remembers where you are in the conversation—and where you are in your data—so you can ask follow-up questions if you want. If you misspeak or if you're misunderstood, you can reword part of your question to the AI agent without having to repeat everything. Also, since Arria Answers knows the context of your question, you need only specify the relevant part of a request, with Arria Answers filling in the unspecified elements automatically.

Convenient. Arria Answers encompasses convenience in two ways:

- Taking simple natural language commands, Arria Answers allows nontechnical users to interact with data wherever it is convenient or whenever it needs to be hands-free (for example, in the board room or while you are driving, in a hospital, in a lab, or in a warehouse setting).
- Arria Answers is easy to use: Its ad hoc conversational format allows users to freely interrogate their data without the help of experts.

"Sales are $1.57 million." To get to the context, to get to the causes, you would have to ask more questions to find out what is going on.

Arria knows what you want to know: "Are my sales as I expect them to be? If not, why?" Because we know this, we can build it into our conversational AI.

Arria's analytics algorithms do the heavy lifting and find actionable insights and important metrics, without you having to specify and look for those insights. Arria Answers encodes the semantics of your domain within the product. This means that Arria Answers is smart enough to know which characteristics make an insight important to you and report that insight in a timely and actionable way.

Because Arria Answers is backed by Arria's deterministic language generation, it can deliver any kind of analysis that you want. The true power of Arria lies in its capability to give voice to your data—your voice, in your way of reporting, with the analyses you choose to apply. But out of the box, Arria Answers delivers variance and key performance analysis. You can effectively see which business units are contributing toward your goals without having to sift through every one manually. You can see what is driving period-on-period change of key measures without having to list each potential driver or offset to see if it has an impact within the period.

PART III

Evolving the Technology

CHAPTER 8

The Collective Mind: The Age of Collaboration

When I was sixteen and evolving as an artist, I decided to teach myself how to use an airbrush. I spent hours in my attic studio practicing my technique—pull the nozzle too much and I would ruin the image, not enough and I wouldn't get the vibrancy I wanted. When I made a mistake, which was often in those early trials, I would tear off the page and start fresh. There were so many days of spraying, tearing, adjusting, spraying, tearing, adjusting. Each time I tore off a page, I thought, *I'm one step closer to perfecting this.* A few faithful friends and family kept some of those pieces throughout the years to keep the memories and lessons alive.

I assume my early perspective was a blend of personality and upbring-ing, but it has always stayed with me. *When I fail, I'm one step closer to succeeding.* In the years since then, I have witnessed how this mindset has served me, both personally and professionally. Each time I make a mistake, I pause to reflect: Am I failing? Or am I one step closer to succeeding? So much of success is tied to our perspective. When you don't like how you're living, change how you're thinking. Whether you call it neurolinguistic programming or fixed versus growth mindset, it's a simple philosophy—once you think differently, then you feel differently, then you behave differently.

There is an old saying, made popular in recent years by author and scientist Jon Kabat-Zinn in *Wherever You Go, There You Are*: "You can't stop the waves, but you can learn to surf." The underpinning of that phrase provides one of the most important truths in life: Change is inevitable. All humans have experienced this on a personal level, and all leaders have felt this on a professional level. Though we often think of change as a disruption to be feared, it is actually a powerful tool for continual evolution.

DIGGING DEEPER INTO MANAGING DISRUPTIONS

From my time as a leader, I have encountered many disruptions that have evolved my leadership mindset. With attention and focus, any disruption can be used as a catalyst for evolution. After all, without disruption, there is no motion.

1. Choose your words.

When a disruption strikes, as a leader your first priority is to mitigate risks, and that means putting your people first. For me, putting my people first means an intentional investment in communication. How do I speak with them? What are the words I choose? How can I authentically express myself to create space for them to do the same? Like all crisis countermeasures, communication is a tool that should be fostered proactively rather than reactively. Be mindful that the words you choose create a ripple effect, not just within your team but within your own psyche.

2. Get focused.

All disruptions can affect your focus and flood your mind with distractions. To counteract this, I intentionally find my focus each day. Where are the opportunities for evolution? Rather than seeing the chaos, I am able to find my calm and envision myself and my team moving forward—not in spite of the crisis but because of the crisis. Knowing how to shift perspectives is one of the most profound tools a leader can use for the betterment of herself and her teams. It's not always easy, but it's an important discipline. Getting focused with the right perspective as quickly as possible clears the mental chatter and clutter that can take up too much mind space. Sometimes you have to empty before you can fill.

3. Embrace your humanity.

When disruption strikes—and it will—you are deeply aware of your humanity. You understand your limitations to control the environment. This can leave you feeling powerless—but only if you let it. Since we don't control all dynamics, we must get comfortable with movement, whether motivated internally or externally. Without disruption, there is no motion. With motion, there is an opportunity to move, grow, pivot, and evolve.

4. Honor the individual, empower the team.

Like all leaders, I want to get the best out of my team. I've learned that in order to do that, I must start with the individual. I don't want to limit anyone's potential. As a leader, my goal is not to control everything; it is instead to move others toward our shared vision. I can only do that when others feel freed to be who they are and bring the totality of themselves to the project. Above all else, honor the individual. You will find that your teams are more productive, efficient, committed, and authentic. There is nothing more powerful than a group of people working with their full potential. What a wonderful privilege for a leader to be able to witness and cultivate that daily.

5. Always double-check your motivations.[77]

I think it's important to take the time to have some self-awareness about what has influenced you, what you stand for, and being conscious of what's actually going on. One of my life principles is to inspect your motivation about why you're doing something. You want to do a self-check that you're on the right path and that your actions aren't driven by ego. That way, you can calibrate your actions and how you communicate around the right purpose. Otherwise, it's easy to get caught up in the moment.

That can be important when you're having difficult conversations.

And those conversations shouldn't be difficult if you've built a relationship of trust, because then the person knows where you're coming from. They already trust you, and they're not feeling defensive, so they don't take things so personally.

---

77 Sharon Daniels, "Sharon Daniels of Arria: Always Double-Check Your Motivations," interview by Adam Bryant, *The New York Times*, January 13, 2017, https://www.nytimes.com/2017/01/13/business/sharon-daniels-arria-motivation.html.

It really comes down to your perspective—how you approach life and what lens you are going to use to look at anything, whether it's a person or a situation. You have a choice to look at somebody's good qualities or their bad qualities, because they're present in every one of us. If you choose to focus on the good qualities in somebody, the outcome will almost always be more positive than when you take the opposite approach.

Know Thyself

The only way to find success, happiness, and fulfillment in a fluid, ever-changing environment is to be firmly rooted in self-awareness. Though there are many resources on building self-awareness, the oldest and most succinct is the ancient Greek aphorism inscribed on the Temple of Apollo at Delphi: Know thyself. This certainly sounds easy enough, but as the world becomes more volatile, it becomes more challenging to achieve. Why? Because awareness requires a pause to self-assess, manage, and analyze.

Of all disciplines, sports training embodies this best. After each competition, athletes have quantifiable results in front of them—times, measurements, and scores. In order to win, they must set aside their insecurities and face the facts before them. They must go deep, be real, and truly hear the feedback from coaches, mentors, trainers, dieticians, and technical coaches. Embracing these hard truths is what separates a good athlete from a great one. Each event offers another opportunity to pause, self-assess, and garner feedback. Athletes wake each day focused on what didn't go well the day before and how they can improve upon it today.

What is the payoff for this radical self-awareness? Evolution. As an athlete, you evolve physically, mentally, spiritually, emotionally, technically. Arria's culture is heavily influenced by one of Arria's board of directors and one of my dearest friends and mentor, Barbara Kendall. Barbara is a five-time Olympian and triple Olympic medalist. During her athletic career in international windsurfing, she has won every color medal, starting with gold in 1992. With her expertise in sports training and leadership coaching, Arria has been able to infuse some of these sports philosophies into the business sphere, including brutal honesty, authenticity, and radical transparency. Barbara's every cell runs on high octane. It is clear when you are in her presence that she is in the flow. Any entrepreneur in the tech space would benefit from having a coach to reinforce and encourage you to know yourself, know your own power, and know the power of a team.

While the CEO of a company generally has the responsibility to define the culture, defining vision, mission, and values doesn't happen by sending a single memo; it happens when you behave, act, and make decisions in ways that support those values on a consistent basis. The vision, mission, and values that define the Arria culture were influenced by the founders of the University of Aberdeen who set out "to found a university, which would be open to all and dedicated to the pursuit of truth in the service of others." We believe that by respecting the cultural ethos and beliefs of the founders of the University of Aberdeen written in 1495, and by honoring their dedication to the betterment of humanity via the evolution of language, we sustain the invisible bonds that enable great institutions to remain in existence through generations, during which time everything around them changes in infinite ways.

Arria's Vision

- Our vision is to ensure, through our natural language technologies, that humanity remains at the heart of AI.
- Empowering people to know more, be more.

Arria's Mission

- Our mission is to lead a global revolution and communication—connecting the datasphere to humanity by giving it the power of language.
- Our mission is to free people to dream, think, create, discover, build, heal, help, relax, live, and love.

Arria's Values

- We believe every word has a spirit. We are one, real, dynamic spirit, unwaveringly committed to our vision and our mission and to taking responsibility for the word.

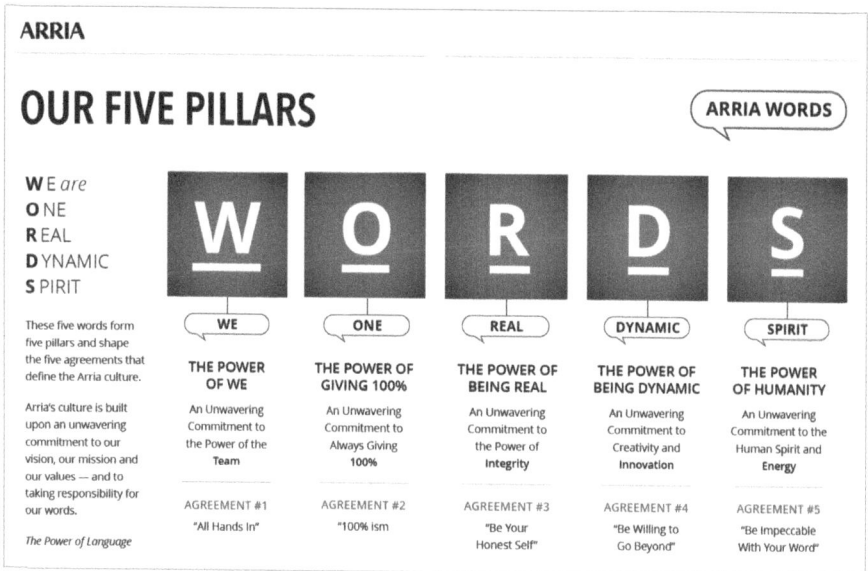

Figure 8.1. These five words form five pillars and shape the five agreements that define the Arria culture.

Whether on the water or in the board room, Barbara and I have both learned that high performance is this continual orientation around who you are, what has improved, and what needs to improve. Knowing who you are is your foundation, and as the environment changes, so do you. In order to continue to know yourself, you must reassess your foundation.

SELF-AWARENESS TRAINING

- What are my values?
- My passions?
- My strengths?
- My weaknesses?
- My motivators?

Once we can answer these questions about who we are today, in this precise moment, we are rooted in our principles, our foundations. With self-awareness, no matter the volatility of the environment, we stand strong. The waves may be bigger than we expected, but we are firmly rooted. We can make decisions easier, pivot quickly, and know that we will be more evolved tomorrow because of the disruptions today.

Competition is incredibly healthy; too much competition, however, is unhealthy. You can certainly win by cheating and unfair tactics, but the most satisfying wins happen by fairness and integrity. If you want to be successful and feel proud of what you've accomplished, you must lead with your personal values. In order to do this, you must bring your whole self to every competition—in the workplace, in the home, in the community. I strongly believe one of the greatest gifts you can give to someone is to have your own act together. The most rewarding thought at the end of life is knowing *I did a good job of being me.*

Interconnectedness

We don't compete, work, or play in a vacuum. We are an interconnected species and world, made more so by the datasphere. How do we leverage this interconnectedness for our own high performance? Through collaboration. Admittedly, there is a fine line between competition and collaboration. The difference often lies in your personal values and those of the people surrounding you. Whether you're meeting a new friend or searching for a new job, you must be aware of the underlying values of the person or organization before you.

In order to evolve, build your world with intention. One of the reasons I love Arria is the similar value sets of our people. Even though we are spread around the world—from creative backgrounds to scientific ones—we share similar principles that underlie our collaborations like a thread that connects us. Though business leaders often focus on operations and strategies, it's these more nebulous principles that create an organization's success and longevity, because it binds people during disruptions. They remain united and rooted in values that orient them in volatile times. These values allow for symbiosis, which, as game developer Nicky Case has pointed out, is nature's "most underappreciated trick."[78]

> Symbiosis shows us you can have fruitful collaborations even if you have different skills or different goals, or even if you are a different species. Symbiosis shows us that the world often isn't zero sum—it doesn't have to be humans versus AI,

---

78 Nicky Case, "How to Become a Centaur," *Journal of Design and Science* (2018), https://doi.org/10.21428/61b2215c.

or humans *versus* centaurs, or humans *versus* other humans.
Symbiosis is two individuals succeeding together not despite,
but because of, their differences. Symbiosis is the +.

At Arria, we embrace a shared value about the symbiosis of humans and AI, which is reflected in our formula: Hu + AI = superhuman. Like Case asserts, humans and AI technology succeed because of their difference. One on its own is not as powerful as the two working in concert.

Shared values, however, are different from shared thinking. Another key to success for any leader, especially those in innovation fields, requires a diversity of thought. In order to innovate, you have to be able to think differently. You need to see what doesn't yet exist. I have the mindset and training of an artist, yet I spend my days surrounded by scientists. If I'm in a room full of creatives, we all might agree with each other. That doesn't propel our collaboration and can impede our innovations. As computer scientist Alan Kay once noted, the entire technological field was originally born from this diversity of thought: "Knowing more than your own field is really helpful in thinking creatively. I've always thought that one of the reasons the 1960s was so interesting is that nobody was a computer scientist back then. Everybody who came into it came into it with lots of other knowledge and interests."[79]

Though our perspectives and thinking are vastly different, the values that underlie our mindsets are essential to our collaboration. Our partnership exists because of our shared foundation—principles of respect, trust, honesty, integrity, growth, and self-awareness. We also

---

79   Alan Kay, "Vannevar Bush Symposium," 1995, https://archive.org/details/XD1941_9_95VannevarBushSymTape10_AlanKay/XD1941_9_95VannevarBushSymTape10_AlanKay.cdr.

must be unwavering believers in the shared vision we hold that allows our innovations to thrive.

Though collaboration and competition are two distinct concepts, they have the most power when combined. Whether you're working to be your highest-performing self or lead the highest-performing organization, you need power. Competition creates this energy; collaboration is the value system that drives it. Pure competitive power must be wholistic. Collaboration without a bit of competition breeds complacency. Competition without collaboration breeds a negative mentality that inhibits creativity. Though the concepts are often pitted together, in a more evolved world, we recognize the paradox that lies within.

Isaac Asimov, biochemist and science fiction author, addresses this paradox in his essay "The Relativity of Wrong".[80]

> In every century people have thought they understood the universe at last, and in every century they were proved to be wrong. It follows that the one thing we can say about our modern "knowledge" is that it is wrong....
>
> When people thought the earth was flat, they were wrong. When people thought the earth was spherical, they were wrong. But if you think that thinking the earth is spherical is just as wrong as thinking the earth is flat, then your view is more wrong than both of them put together.

The basic trouble, you see, is that people think that "right" and "wrong" are absolute, that everything that isn't perfectly and completely right is totally and equally wrong.

---

80   Isaac Asimov, "The Relativity of Wrong," *Skeptical Inquirer* 14, no. 1 (Fall 1989): 35–44.

Allowing for diversity of thought doesn't just apply to a team; it applies to the individual as well. When we can step outside of "right" and "wrong" as Asimov suggests, we open ourselves to a broader world of nuance. This allows us the space we need for original thinking and freedom of expression. Living this as individuals allows us to model this for our teams, which in turn creates environments where creativity thrives and innovation reigns.

Finding the Zone

Understanding the balance between collaboration and competition allows us to find what athletes call the zone—that place of optimum performance. When we're in the zone, we feel engaged, purposeful, alive, human. For Barbara, she found that zone in competitive sports. For me, I found it in building businesses.

Many people who have felt this flow of energy will spend decades trying to rediscover it. In my own personal and professional work, I've found the best technique to replicate the zone is to cultivate continual self-awareness. When I can pause, reflect, and garner feedback from trusted sources, then I am investing in my own continual evolution. I can accept and embrace that I'm not the same day to day, month to month, year to year.

No matter our background, we all benefit from an athlete's mindset. Having an orientation toward self-awareness then allows for self-leadership—a readjustment of thought, perspective, training, or action when required. Just as an athlete propels their own continual evolution, so can all humans. We remain empowered, regardless of the fluid, volatile environment around us. After all, the one thing we can count on is change.

We don't control the environment, but we do control our perspective, our evolution. Our technological innovations can and should support our evolution. As Charles Darwin noted in his *On the Origin of Species*, it's not the strongest of the species that survive, or the most intelligent, but the one most responsive to change. I would say this applies to technologies as well.

Arria's culture is built on the power of language: We believe our humanity is defined by language. Since every word has a spirit, we feel it is our responsibility to care for the spirit of the word. Our culture is built upon an unwavering commitment to our vision, our mission, and our values—and to taking responsibility for our words.

To this end, our Arria team members have chosen words that resonated with them and their unique perspectives on the world. I chose the word "vision" because if the vision is big enough to create such passion and enthusiasm in the team working to realize that vision, nothing can stand in its way. I love being part of a team filled with the energy that comes from collaboration and being part of something bigger than ourselves.

At Arria, our vision is to ensure, through our natural language technologies, that humanity remains at the heart of AI. Empowering people to know more, be more. Admittedly, we have a very big vision. In fact, it's so big you could ask, "Do we have a vision, or does the vision have us?" I think it is the latter. That's my perspective, and each day I'm one step closer to achieving it.

NOTES FROM AN UNWAVERING BELIEVER

Barbara Kendall, Five-Time Olympian and Triple Olympic Medalist

In my early sports career in windsurfing, I was fortunate to have supportive leadership that modeled a values-centric approach. Those values reflected down to my teammates, and as such, we enjoyed collaborative fair play. We all trained together. We weren't competitors; we were collaborators. The only way to get good individually was to work together. We shared the same view of competition and pushed ourselves and each other to perform at our full potential. We were only as good and as evolved as the people around us.

When I got to the Olympic games, however, I saw the dark side of competition. After all, there is only one gold medal, one silver, and one bronze. Governments pour so much money into their athletes that it can get nasty. It didn't take long for me to understand there are two types of winning: one achieved by cheating and one by fair play. Though both ways might come with a medal, they are born from disparate values. When you win fairly, through a principled approach, it feels differently. There is honor, pride, satisfaction. You find that others lift you up and share the win with you.

In order to find the balance between competition and collaboration, you find others who reflect similar values as your own. You maintain a principled foundation. This doesn't mean you are passive. It means you still fight and bring your all, but you lead with grace and integrity. This way, no matter who wins, you are all supported and, in that sense, you all win.

At the start of any competition or collaboration, you think of the end result: not the medals but the game. You think about how you want to feel when it's over. How you want others to think of your game and your character. Then, you must bring your whole self to the game.

Though there are many ways of optimizing performance, they all start with a willingness to be open. Open to what? To seeing other perspectives, to collaboration, to competing with integrity, to feedback. The goal is to avoid being static. Each practice is different from the day before, because I'm different from who I was the day before. Each time I compete, I'm different from who I was before. From one Olympics to the next, I was a completely different person—not just physically but emotionally and spiritually. I couldn't reuse the same formula that won me a gold medal the first time because it wouldn't work again. That is intentional. As an athlete, my goal is constant evolution, and that's a lesson that serves me in life and in career.

One of the reasons we avoid being static is because the environment itself is fluid, so being static doesn't serve me at all. With windsurfing, the environment is constantly changing. There is little reason to practice a static formula because it won't serve me in a fluid environment. Instead, I have to build a principled foundation so that I can rely on my intuition during competitions. If I had not trained in that way, I would have been left behind. Though I learned these lessons in sports environments, they continue to serve me, and those I train, in professional and personal environments as well.

CHAPTER 9

The Future Is Yesterday: The Great Cycle of Technology-Driven Life

As a young girl, I remember the regular visits from my parents' burly accountant. He would arrive with a large bag and crowd our kitchen table with stacks of papers and his whirring adding machine. Once he got set up, we'd hear the familiar cadence of his work—*dat dat dat thrum dat dat dat thrum dat dat dat thrum*. It would stick in my head like a song for the rest of the day. To me, that calculating machine was a marvel—and it truly was. What would have taken him hours of calculating—and double-checking—could be completed in an hour. Since that time, we have seen the invention of calculators, Excel, and other technologies that can not only cull through mountains of data at a higher rate but can now turn data into BI that can communicate in language, not just numbers or pictures.

Because I'm of the generation that witnessed the birth of so many of the innovations that many take for granted, it is all miraculous to me! Even though these technologies have become so ingrained in our daily lives that we don't even marvel at them anymore, many innovators like myself continue to look to the past to pave the way to the future. Why? Because we can learn from the cycles of adoption we've witnessed before. We can even sometimes recognize human needs based on the technologies that fade and those that endure.

By definition, innovators are moving things forward and pushing us to evolve. According to software engineer and former CEO and chairman of Google, Eric Schmidt, this change is not consequential but is essential. In fact, he considers flexibility as something to be leveraged by innovators: "Technology is always evolving, and companies ... not just search companies ... can't be afraid to take advantage of change."[81] In my career, I've learned to get comfortable with volatility, disruption, and chaos. Not only is it something Arria can leverage in our industry, but it's also something that helps us evolve collectively and as individuals.

As a creative person, I have always been interested in where ideas come from. What is it that makes some innovations successful and others not? How is it that some people can see around the corner to what humans *might* need, *might* want, *might* require? And how do they stay committed to those ideas even when others might doubt them?

Having witnessed some of the most advanced innovations of the modern age, I've noticed a pattern when bringing something new to the world. Ruth Stafford Peale, wife of Norman Vincent Peale, coined the phrase, "Find a need and fill it." As innovators, that's exactly what

---

81   Dr. Joseph Aluya and Dr. Ossian Garraway, *The Influences of Big Data Analytics: Is Big Data a Disruptive Technology?* (AuthorHouse, 2014), 140.

we try to do—though the finding might come easy, the filling often does not. This requires a focus on the future, and it also requires a vision of the past. In order to innovate, you must first identify a need. This is a current need or an envisioned future one. This is where the idealized notion of innovation comes from and is what we imagine the visionaries of our time do—think about the future. In reality, however, some of the most brilliant innovators I've worked with also understand the power of the past. It is by looking to the past that we can identify gaps. What was missing? How could it have been better?

BRINGING SOMETHING NEW TO THE WORLD

1. Identify need:
   - What's there?
   - How are we currently doing something?
2. Identify what's missing:
   - How could it be better?
   - What's the value of making it better?
   - What are the pain points if it's not made better?

This is admittedly a simplified equation for innovation. The truth is that it's difficult to bring something brand new to the world. As Peter Thiel explores in *Zero to One*, it's easier to duplicate what already exists rather than create something entirely new. Regardless of how complicated, Thiel asserts that innovation is essential to humankind:

"Humans are distinguished from other species by our ability to work miracles. We call these miracles *technology*."[82]

Though they may be miraculous in the eyes of unwavering believers like Thiel, I've learned in my thirty-year career of building and expanding technology start-ups that it's easy to get infatuated with your own idea—your own "miracle"—because you recognize how it can improve the lives of others. Every innovator dreams of creating the next newest, biggest thing on the market—the proverbial unicorn in a field of horses. It is certainly a magical feeling, but it's shortsighted not to acknowledge the inevitable patterns that arise when you bring new things to market. In fact, much of an innovator's success is tied to their understanding of human behavior and their preparation for the evolution of innovation.

THE EVOLUTION OF INNOVATION

1. Disruption
2. Discovery
3. Universal

When an innovation is new, no matter how needed and anticipated, it is a disruption to the norm. Even though innovators might be infatuated with the value of their breakthrough capabilities, we can't escape a human truth: Unfamiliarity breeds fear. Even something as useful and innocuous as a calculating machine initially breeds distrust. You have to allow for that adoption curve to take place. It needs to be built into your business plan because new things are harder to bring

---

82   Peter Thiel, *Zero to One* (Crown Business, 2014), 2.

to market. This shouldn't discourage innovators. Rather, it should be prepared for and embraced as a by-product of the innovative process.

After the initial disruption phase, there is a period of discovery when users experiment with ways the technology might benefit them. This is an exciting moment for the innovators because the product is now in the hands of users. They can objectively see how it's used and whether it is easing the pain points as intended. Over time, and with increasing numbers of users engaging with it more frequently, the technology becomes less novel—and therefore less scary—and we see how it makes our lives easier. In the final phase, the technology is no longer deemed a threat. Users embrace the ways it makes their lives easier, and it is adopted universally.

I have witnessed this cycle countless times in my career—from introducing new business strategies to new technologies—and my time at Arria has been no different. Some innovations, like natural language technology, generative AI, and agentic AI, are ahead of their time and need an incubation period to wait for other technologies to catch up and for early adopters to need—and ready themselves for—its benefits. When we were introduced to natural language generation in 2011, we could see what was around the corner. As the world digitized everything, we at Arria recognized the pressing need to aggregate data. After all, what good is information if you can't extrapolate meaning from it?

Nevertheless, it took ten years-plus before the adoption rate kicked in. Still, we held on to our vision. Soon, early Arria adopters discovered the value of NLTs to bring the human way of communicating and understanding to a business platform. They recognized that when we can replicate the way people think, we can take that knowledge and share it widely. For experts, we can capture the learning and experiences amassed over decades of discovery; for nonexperts, we can offer expert capabilities at the touch of a finger. Adopted universally, we can empower individuals, break down silos, and enable expertise to go broader and wider across an enterprise or the world at large.

We often think technology is separate from humanity, but it's not. If we can move past the initial fear-based stage, then technology can evolve us further and make us more human. Just as my parents' accountant learned when he embraced the newfangled adding machine, it freed him to spend less time at our kitchen table, double-checking his calculations, and spend more time doing what made him uniquely human—connecting with others, helping grow their businesses, and cultivating his own joy and purpose. This is the paradox that we always return to: Technology is an inflection point that propels us into a greater, more evolved human species.

A Kodak Moment

Technology has been allowing humans to evolve since the wheel. There has never been a time when humans haven't been inventing and deploying some form of technology to make life easier. Innovation isn't new, but digitization is. Digitization involves removing physical constraints and limitations. If we look to the past, we better understand where the future of digitization is headed. More specifically, the history of the digital photo foretells the future of printed reports.

The first cameras were 8 pixels, followed by 16 pixels, then 32 pixels, 64 pixels, and 128 pixels. In short, it took twenty years to get enough pixels to take a decent picture. In the same way, Arria's science team has been building natural language technologies for over thirty years, and Arria has been commercially automating knowledge work for six years, yet few still understand its pending impact. If the business community is Kodak, then natural language automation is the digital camera. Arria is the missing layer that allows all other technologies to interact with humans in real time in natural language. As we learned with the digital photo, when something valuable can now be freely accessed in digital form, everybody can have it, and the cost of reporting production is minimized or eliminated completely.

People take 900 billion more photographs today than they did in 2000 because the barriers to producing them have been eliminated due to digitization. When expertise is digitized, the value placed on analytics, for example, will move from high-cost human form to low-cost digital form.

In essence, just like anybody can now take a high-quality digital photo and edit it, anybody can afford to have the consultation of an expert because that expertise is now free of cost.

There are now 1.3 billion high-quality cameras being produced and sold annually at minimal price, and the number of photographs taken per year has increased by 900 billion. Why? Because it's free. It's effortless. The results are fantastic. How many reports and narratives will be produced when they are instant and free? How many financial reports will be created instantly rather than quarterly when the expertise necessary to write them is digitized?

It's only by looking back that we see the path we walked, but it is only by following that path forward that we evolve. If we're using the past as a model for the future, then we recognize that the availability of expertise on demand is predicted to revolutionize human existence. And if we doubt that, we can reach into our pockets and pull out our phones. We are having a Kodak moment, but the whole world is Kodak and Arria is its digital camera. The world is waking up in stages to the reality that the missing layer in the current technology stack is the power of language. Perhaps we are all like Kodak and cannot see that the thing we are best at is now digitized.

Looking at how the digital photo has evolved over time helps us understand the future of language technologies. We can see how digitization removes barriers and empowers users with expertise that is democratized and on demand. If we don't recognize how this can empower human existence, then we merely need to think about the last time we took a roll of film to be developed. Now, we merely take our phones from our pockets and hold the power. The same is true for generative AI technologies that take the expertise of many and share it with all.

The Role of Innovator

Kevin Kelly, author and visionary founder, champions technology's values and impacts. He holds that humanity itself is the greatest invention, and the greatest benefit of technology is its ability to allow us to continually reinvent ourselves: "Technology is more than just the stuff in your pocket; it's more than just gadgets; it's more than just things that people invent. It's actually part of a very long story—a great story—that began billions of years ago. It's moving through us, this self-organization, and we're extending and accelerating it, and we can be part of it by aligning the technology that we make with it."[83]

Being an innovator is not just about identifying needs; it's also about holding a vision. Innovators must be their own unwavering believers in the value they offer to others while also embracing the ways that initial vision may morph and shape. Sound paradoxical? It is. In fact,

---

83   Kevin Kelly, "Technology's Epic Story," TED Talk, November 2009, https://www.ted.com/talks/kevin_kelly_technology_s_epic_story.

it's all a paradox. Our idealized view of innovators is that they have their eyes locked ahead on the horizon, and they certainly do. What is perhaps less romantic but equally essential is also having their feet firmly rooted in the past and ultimately the present moment. We can't ease pain points, for example, by looking ahead. It's not the pain points of future generations we are concerned with but those of past and current ones. What can improve for our families, our colleagues, our communities, our world *now*? Using the pain of the past to inform the pleasures of the future is the great cycle of technology-driven life.

Dennis Gabor, awarded a Nobel Prize in physics, once wrote:[84]

> We are still the masters of our fate. Rational thinking, even assisted by any conceivable electronic computers, cannot predict the future. All it can do is to map out the probability space as it appears at the present and which will be different tomorrow when one of the infinities of possible states will have materialized. Technological and social inventions are broadening this probability space all the time; it is now incomparably larger than it was before the industrial revolution—for good or for evil.

The future cannot be predicted, but futures can be invented. It was man's ability to invent which has made human society what it is. The mental processes of inventions are still mysterious. They are rational but not logical, that is to say not deductive.

What we know about bringing new products, services, and technologies to market is that they all require one thing: the human mind. When we realize that all innovation is centered on easing our pain points and improving life's experiences, we begin to understand that technologies exist because of human creation. Therein lies another

---

84   Dennis Gabor, *Inventing the Future* (Alfred A. Knopf, 1964), 207.

paradox: Human nature is the essential core of AI. One doesn't exist without the other. Though the two are often pitted against one another, they are, in fact, two fibers of the same creative thread. All AI innovation—as mysterious as it is—is powered by one indomitable force: the human spirit. This is true of our past, our present, and our superhuman future.

CHAPTER 10

Building the Legacy of Language

Humanity is defined by language. Language is so automatic to us that we rarely notice that we even use it when we're thinking. When we speak, we pull from the full knowledge of our experiences—from what we've observed, learned, felt, and remembered. In an instant, we analyze, calculate, and then communicate. AI is technology that's trying to replicate that human intelligence. Imagine trying to do that! Even though we take it for granted, it's a miraculous process that has allowed us to be the most advanced species on earth.

We're born with the capability of language, but what does that mean? It means we're able to receive inputs and to "know" what these inputs mean. We're able to analyze the meaning of the inputs, not only in and of themselves but also in relation to other pertinent bodies of knowledge. And then we're able to communicate the conclusions of this cognitive process using words that can be understood by the listener. It is this process we call "language," and it is, of course, the primary function of the human mind.

By age one, we have roughly five hundred words in our vocabulary. By the time we're ten, we have about five thousand. By the time we're fifteen, we have about fifteen thousand. If we continue advancing or go to college, for example, we increase that vocabulary by ten thousand to twenty thousand words. If we continue to specialize, we can increase our vocabularies to fifty thousand to eighty thousand words.

If a set of identical twins is raised on a farm with only their parents, they will essentially share an almost identical number of words. If one moves to the city and studies a high-level specialty like heart surgery, for example, when she returns home, she will have thirty thousand to forty thousand more words available than her twin. Even though the twin is the same person she was before moving away, she is a different construct because her ability to describe phenomena is infinitely greater just by her broadening her sphere and specializing in a field.

Since there are so many words, there is literally an infinite number of ways to describe something. Let's consider a sunrise. When the sun rises, a sensor could detect the date and time and record those numbers in a column. That's one way to capture the sunrise. A person, however, might wake and capture the moment in words: *Look at how the sun strikes the dew drops on the daffodils.* Another might say, *After my late night, the sunrise was too bright and made my head ache and my stomach churn.* And still another might reminisce: *Today's sunrise reminds me of my family trip to Hawaii in 2017.* One simple sunrise can trigger an infinite number of experiences. What defines each experience is its expression through the power of language.

At our essence, a human being is defined by consciousness and the power of language that is so innately human that when we see it happening outside of ourselves, we fail to recognize the mind-blowingly esoteric, unexplainable significance of what's happening. Since we're born with the capability of language, when we see it literally being transferred to the machine, we just yawn. In fact, when we showcase NLTs, we often advise viewers not to blink, or they will miss the magic moment. If I take a data set and turn it into a twenty-page narrative, for example, this transformation happens as quickly as I can hit a key. Blink. It's over. My advice? Don't blink.

The History of the Datasphere

We live in a world of data. Data is a language, but one that does not involve words. Though we often don't recognize its impact because it's become so much a part of our lives, data helps us measure, monitor, and improve things. Technology has been allowing humans to evolve since the wheel. As long as there has been human life, there has been some form of technology being invented and deployed to make life easier.

If we consider the evolution of the mobile phone, we understand how these innovations transition into our daily lives. In 2007, for example, there were zero iPhone users; in 2016, there were 1 billion. In fact, the iPhone 4 has the same compute power as the most expensive supercomputer in the world in 1975. And now we carry them in our pockets like it's not miraculous! Such innovations had massive impacts on the digitization of our society, and specifically the internet.

What we once called the "internet" has evolved. Just as the word "phone" no longer adequately describes the power of a smartphone, the word "internet" no longer describes the multifunctionality of the datasphere. The datasphere is the internet and everything that's connected to it. As you can imagine, this is difficult to quantify. What are the devices connected to it? What are the things connected to it? What is the scale of it? What is its physical size?

The datasphere is the multidimensional reality of the internet. What began as one web page in 1993 has now exponentially grown to over 1.1 billion webpages in 2025.[85] Not only has the volume grown, but its functionality has evolved. At Arria, we have determined that the datasphere's physical dimensions are created by its compute power, social flow, commercial flow, connected devices, connected things, and overall scale.

TYPES OF CONNECTED DEVICES:

- Personal computers
- Smartphones
- Tablets
- Watches
- Servers
- Server farms/data centers
- Mainframes
- Supercomputers
- Closed-captioned televisions
- Automatic teller machines
- Credit card machines
- Number of routers/modems
- Number of satellites (internet/phone)
- Number of databases

85   Nasir Uddin, "How Many Websites Are There on the Internet in 2025?," MuseMind.agency, updated May 5, 2025, https://musemind.agency/blog/how-many-websites-are-there.

The datasphere began as the internet of documents. Next, with the inclusion of transactions, it became the internet of commerce. With the addition of applications that allowed us to do things, like email, we added a new function. In 2004, it became the social internet. More recently, it became the internet of things with a myriad of devices connected to it. Over time, it grew until it became a repository of experiences.

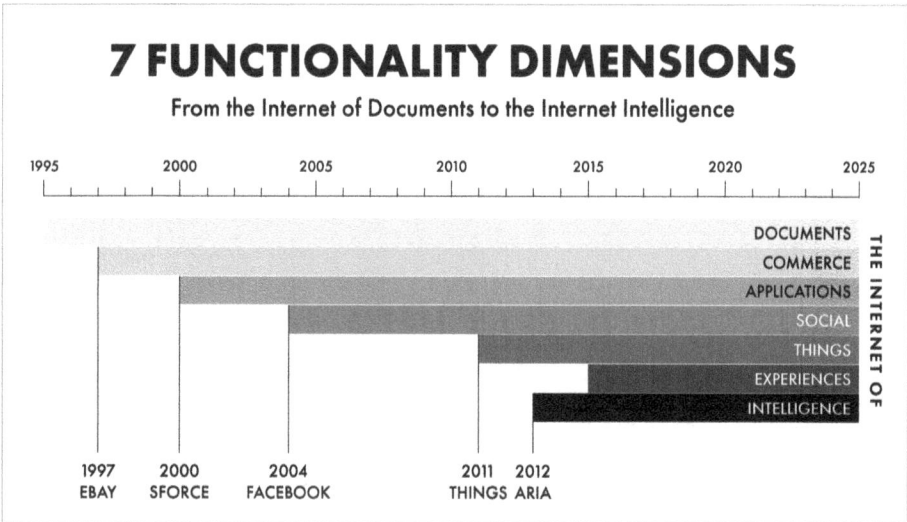

7 FUNCTIONALITY DIMENSIONS
From the Internet of Documents to the Internet Intelligence

1995	2000	2005	2010	2015	2020	2025

DOCUMENTS
COMMERCE
APPLICATIONS
SOCIAL
THINGS
EXPERIENCES
INTELLIGENCE

THE INTERNET OF

1997	2000	2004	2011	2012
EBAY	SFORCE	FACEBOOK	THINGS	ARIA

Figure 10.1. The evolution from the internet of documents to the internet of intelligence

Despite these evolving functionality dimensions, however, the datasphere has remained a parrot; it can only respond based on its programming. In short, there's no intelligence in there. It can't put sentences together. It can't have a dynamic conversation. Let's consider this familiar exchange. If I say, "Hey, Siri, what's the temperature in Chicago?" I will hear, "It's fifty-five degrees right now in Chicago, Illinois." I follow up by saying, "Hey, Siri, what will the temperature be tomorrow?" The response might be, "The high tomorrow will be seventy degrees." Though it is an amazing feat that my phone just answered my questions, it did not have the ability to understand that my follow-up question about tomorrow's weather was also about Chicago. Instead, it defaulted to my current GPS location.

Though a technological achievement indeed, this is also an example of a failure of multiturn conversation. It's not articulate. With the advent of natural language technologies, for the first time in human history, the power of language has now been captured in software and has been automated, moving outside of the human mind. The advent of natural language automation changes everything. Why? Because beforehand, the internet of things was a parrot without the power of language. With the advent of generative AI, we have taken the internet of information and enabled the internet of intelligence.

DIGGING DEEPER INTO MULTITURN CONVERSATION WITH ARRIA ANSWERS

Arria Answers is an advanced conversational AI application that brings together natural language automation and insight analytics and turns structured data into intelligent narratives. By replicating the human process, it instantly helps identify, understand, and communicate key insights. No code required! It's a simpler, faster way to analyze data.

If you're looking at a dashboard, for example, and have some questions about the underlying data that haven't been answered in the accompanying narrative itself, now you have the opportunity to have a conversation about the data. For example:

Speaker: What are my underlying sales?

Arria Answers: Overall actual sales are $111.29 million.

S: What are my actual sales in Canada?

AA: Overall, actual sales in Canada are $2.3 million.

S: How have my actual sales changed year-over-year?

AA: Actual sales decreased by $33.07 million (23 percent) this year compared to last year.

S: What drove actual sales versus target sales for Company A in Canada?

AA: Canadian actual sales for Company A in Canada are higher than target sales by $1.44 million (10 percent) driven by brand store, departmental, and channel partners and offset by online and chemists.

Arria Answers's out-of-the-box narratives are user configurable and can be based on visuals or all underlying data. You can also choose narrative length, from top-line summary to narratives that drill down to explain what and why.

Arria narratives deliver contextualized summaries, reducing the number of drill downs needed—and therefore reducing analysis time from hours to seconds. Speed to information means you can impact operational efficiencies faster. It also removes inconsistency of data analysis (a.k.a. human error) because Arria is 100 percent accurate, 100 percent of the time.

Arria has the privilege and responsibility of bringing one of the most advanced AI capabilities the world has ever seen. Whenever anyone, anywhere, is interacting with the datasphere and wants real-time, expert, personalized information delivered to their device in plain English, Arria will be the technology platform that delivers it to them.

Think of the Arria technology platform as a new form of mind for man—the digital mind. But in order for such a mind to coexist in our world, to enable it to connect with other minds, such a mind must have a "body," a host organism. Arria, along with the natural language infrastructure that it is building, is becoming that host organism. Essentially, we're taking language from the mind of man and transferring it piece by piece to the digital mind.

We envision one day soon, Arria will

- contain the collective "mind," the collective expertise, of all mankind;
- maintain live interactive connections with all other minds; and
- maintain all that will thus pour through it.

To be capable of such a task, Arria and others must be beautifully designed and lovingly created so that it is strong enough to bear such a load, to bear such a responsibility. It must also be dynamic enough to be able to evolve and grow so that it can absorb anything that may be asked of it—including things as yet unknown.

In 1993, the human mind was full of expertise, and the internet was empty of expertise. With time, that knowledge shifted to the internet, but it wasn't until natural language automation that humans gained access to the subject matter expertise of the datasphere. With Arria technologies, language sits on top of the datasphere. It can say in a single word—yes or no—what might have taken a million data points to inform. As it becomes more interconnected, and more analytics are put into the system, it becomes more advanced and more articulate.

The digital mind is the datasphere—the internet of everything—including our connectedness to it. We're all extensions of the digital mind. In fact, the digital mind doesn't exist without the human mind. The two will continue their symbiotic relationship. That's why it's our mission to keep the power of language at the heart of AI. We welcome the augmentation the digital mind brings to the human mind, and we resolve to become more educated, more empathetic, more understanding, more evolved as a result. What has motivated us from the beginning and continues to do so today is *to be open to all and dedicated to the pursuit of truth in service of others.*

Our mission is to lead a global revolution in communication—connecting the datasphere to humanity by giving it the power of language. As Google leads in search, as Meta leads in community, as Microsoft leads in enterprise, and as Amazon leads in commerce, our goal is to lead in data communication. We want to free people to dream, think, create, discover, build, heal, help, relax, live, and love. For the first time since time itself began, the very thing that defines our humanity, the very thing that makes us the most advanced species on earth—the power of language—has been digitized and moved outside of the human body. Having access to this shared intelligence through human language frees us from robotic tasks and empowers us with superhuman potential. The next great step in human evolution is powered by technology and enabled by language.

Conclusion

What is the human species without language? It's so engrained in how we think and communicate that we rarely think about where—and if—we would be without it. Nelson Mandela said, "Without language one cannot talk to people and understand them; one cannot share their hopes and aspirations, grasp their history, appreciate their poetry, or savor their songs."[86] Language is the ultimate form of communication, and as such it unites us, divides us, comforts us, guides us, inspires us. Whether we are considering the words we choose to communicate insightful meaning within data, the words that we choose to use with one another, or the words we choose to describe how our life unfolds, in each instance the words have power.

It's becoming more and more relevant in the technological and human worlds that there is space for everyone's perspective. We don't need to change anyone's mind; we need to expand our own. In fact, differing perspectives fuel personal evolution in life and in business. Most leaders have found they have higher functioning, more efficiency, and more creative teams when there is diversity of thought. The same is true on an individual level. One of the reasons Arria excites me is because I get to play with analytics and language. It's quite a juxtaposition that uses

---

86   Nelson Mandela, *Long Walk to Freedom: The Autobiography of Nelson Mandela* (QMB, 1995), 84.

both sides of my brain. I've found that when I can use the totality of my thinking, I am my most productive and creative. The world opens up to me, and I can maintain the broadest perspective.

The same can be true of innovation. What we're discovering at Arria is that, when you combine human experience with technology expertise, magic happens. When you can capture knowledge, then you can replicate it and share it. After all, insights are great, but answers are better. Natural language technology is a universal capability that analyzes and thinks like Einstein and speaks like Shakespeare. No matter how you want to talk about something, whether it's around financial information or retail information, NLT has made it possible. We continue to work to convert data into language that's meaningful, contextualized, and relevant to humans. Why do we do it? Because we want humanity kept at the heart of AI. We want language to continue to aid our personal and communal evolutions. In helping one human, we help all.

When we realize that all is a paradox, we are free to *be*. Technology is made to be binary; humans, however, are not. We need believers and doubters, for one is meaningless without the other. Our competitor is also our collaborator. There is no you and me.

Looking at one single event, there are an infinite number of perspectives you can take. If you don't like what you see, visualize a dial that you keep turning until you find something that's pleasing to your being and productivity. Though we all experience disruptions—and sometimes global disruptions that affect us all—these individual and collective traumas bind us and move us closer to a more evolved species. A different perspective is to consider that we live in the most sustained period of peace in the history of humankind. In fact, the average number of deaths by war has plummeted, the average life expectancy of humans globally has doubled, and there is a significant rise in the leisure class.

PEACE ON EARTH:[87]

- The average number of deaths per 100,000 people on earth by war has plummeted.
- In the period 1905–1929, the number of deaths by war was twenty per 100,000.
- From 1932 to 1952, it was fifty.
- From 1956 to 1980, it was less than two.
- For the past forty years, the number of deaths per 100,000 by war was less than one.

LIFE EXPECTANCY:[88]

- Since 1900, the average life expectancy of humans globally has doubled from thirty-five to seventy years.
- In the UK, in 1850, the average life expectancy was thirty-nine years.
- In 1900, it was fifty years.
- In 2011, it was eighty-one years.

LEISURE CLASS:[89]

- In 1950 there were 25 million international tourist arrivals globally.
- In 2016 there were 1.3 billion international tourist arrivals globally.

---

87 Max Roser, Joe Hasell, Bastian Herre, and Bobbie Macdonald, "War and Peace," 2016, OurWorldInData.org, https://ourworldindata.org/war-and-peace.

88 Max Roser, Esteban Ortiz-Ospina, and Hannah Ritchie, "Life Expectancy," OurWorld-InData.org, 2013, https://ourworldindata.org/life-expectancy.

89 Max Roser, "Tourism," OurWorldInData.org, 2017, https://ourworldindata.org/tourism.

As Chris Kutarna, coauthor of *Age of Discovery*, wrote, "The fragility of this age confronts us daily in our newspapers and our news feeds. And it is far more urgent, more real to us than that other, equal truth. What was it again? Oh, Yes: *Every other generation that has ever lived wishes that it could be alive in ours.*"[90] So what are we doing with this better and longer life? At Arria, it is our hope that because of recent technological innovations, we are spending more time doing things that make us uniquely human—like dreaming, creating, discovering, feeling, healing, helping, relaxing, living, and loving. The greatest renaissance in human history is already moving at full speed, and we at Arria are about to move it to warp speed. We are unleashing a life-altering, game-changing technology that will change the future for the better.

To be responsible for leading a team dedicated to continuing the evolution of language is a stewardship, a responsibility, an advancement that will impact the world in ways we are yet to imagine. Digitizing natural language is the pinnacle of AI. What could be higher? The power of language, the very thing that makes us uniquely human, is moving from the mind to the machine. While historians will record the primary impact of what we are doing (not in financial terms) as a critical inflection point in the evolution of language (and thus humanity), its impact will nevertheless be measured in trillions, not billions, of dollars. In fact, McKinsey reports this automation category as having a $2.6–4.4 trillion impact.[91]

I never waiver from understanding the magnitude of what we are doing as we move the power of language from the mind of man to the machine via our NLTs. I know the mission we are charged with is

---

90 Chris Kutarna, "Our New Renaissance," *Vogue*, January 1, 2017.

91 "The Economic Potential of Generative AI," McKinsey Global Institute, June 2023, 3.

a quest—a long game. Language and humanity will continue to grow and evolve forever, and now we will be communicating with machines as part of that incredible evolutionary future.

It is essential to understand that our language technology did not just pop out of nowhere. It was developed in the computer sciences lab of the University of Aberdeen and in many respects a consequence of over five hundred years of dedication to language that began with the mission: "To found a university which would be open to all and dedicated to the pursuit of truth in the service of others." This powerful intention that initiated the university's dedication to language, that propelled it through over five hundred years of progress, informs us and inspires us today.

The emergence of advancements in language technologies from its computer sciences lab was the logical next chapter in the story of language observed, recorded, and celebrated by the University of Aberdeen since its founding in 1495. It is this technology that Arria has been improving and developing. In truth, Arria is at the vanguard of the process of connecting humanity to the datasphere by giving it the power of language.

Technology has evolved; now it's our turn. The human mind is still the most advanced computer. Now, technology is trying to keep up with the human brain, so the human being has to keep up with technology. Like all things, it's a perfect paradox. If we allow technology to assume some of our rote behaviors and tasks, what can that free us to do?

Data, like language, is meaningless if there is no receiver of the information. If you give a speech, but there is no audience, then it's not a complete process. We need the speaker and the listener, the signified and the signifier. Data is the same way. After all, what is the value of machine learning if it cannot communicate what it learns?

Indeed, this is an extraordinary moment in human evolution. And it is extraordinarily fragile. We are called to action by the historic stakes of the time we live in. To welcome, not suppress, genius. To make new maps to help us better navigate a changing world. What we do in this moment is up to each of us to decide.

We are in the right place at the right time. How we view it is up to us. I remain an unwavering believer in the future potential of humans and technology. From this point forward, nothing will be the same. Our future will be changed for the better in ways we cannot even imagine. Cast fear aside, crank the dial, welcome genius, and embrace the evolution.

About the Author

Sharon Daniels is a member of the shareholder group that founded Arria NLG. Sharon has over thirty years of experience in strategic business development and has consulted with leading global technology and financial service companies. Sharon is also a cofounder of Diligent Corporation and served on the Diligent board through 2010, where she also held an executive position as chief marketing officer.

Prior to joining Diligent, Sharon was a vice president at the US-based company PaperDirect, Inc., where she was instrumental in helping grow the business from a start-up to over $100 million in sales in less than four years. Getting comfortable with computers and technology began in the early '80s when Sharon cofounded Computer Discount of America, a pioneer in the computer mail order industry, at the early age of twenty-six.

Sharon is chairman and CEO of Arria NLG Limited. She also chairs Arria's Global Advisory Council of respected leaders who bring subject matter expertise from multiple industries.

Sharon is a Forbes Technology Council member and has published articles including "How NLG Technology Can Provide Comfort in a Changing World," and "Giving Machines the Power of Language While Keeping Humanity at the Heart of AI." Sharon has been featured in multiple industry publications as a thought leader, including *The New York Times*.